THE DINGO

IN AUSTRALIA AND ASIA

THE DINGO
IN AUSTRALIA AND ASIA

Lawrence K. Corbett

COMSTOCK/CORNELL
Ithaca and London

First published 1995 by Cornell University Press.

Library of Congress Cataloging-in-Publication data

Corbett, L. K.
The dingo: in Australia and Asia / Lawrence K. Corbett.
p. cm.
Includes bibliographical references (p.) and index.
ISBN 0-8014-8264-X
1. Dingo. 2. Dingo – Australia. 3. Dingo – Asia. I. Title.
QL737.C22C65 1995
599.74'442–dc20 94-33541
CIP

Available in Australia and New Zealand through:
University of New South Wales Press Ltd
Sydney 2052 Australia
Telelephone (612) 398 8900
Fax (612) 398 3408

Printed in Australia by Southwood Press Pty Ltd

FOREWORD

Dingoes arrived in Australia about 4,000 years ago. European settlement began just over 200 years ago. Even today we still don't know how the dingo fits into the jigsaw of wildlife interaction and its relationship with the traditional settlers of Australia.

The dingoes' relationship with European settlers has always been subject to mistrust and misunderstanding. Laurie Corbett's compendium of dingo observations is a must for any sheep and cattle grazier who wishes to co-exist rather than fight the dingo.

The Meat Research Corporation is pleased to have been associated with funding some of the research mentioned in this book.

Robin J Bligh
Chairman
Meat Research Corporation

PREFACE

Yellow dog dingo
How old are you?
Where do you come from?
What do you do?

Such questions, posed in a song by the Australian composer, Larry King, have long been bandied about by scientist and layman alike; yet the myths and misconceptions surrounding the origin and prowess of Australia's native dog remain. This book examines the past, present and future status of dingoes. It integrates results from research over the past three decades to dispel the myths and to place into perspective the ecological role of dingoes in natural and disturbed (pastoral/agricultural) ecosystems. Dingoes are also compared with other canids (wild dogs).

This book is divided into ten chapters. Supplementary information is presented in boxes within chapters. Chapter 1 describes the origin, ancestry and world distribution of dingoes. If not with Aborigines, how did dingoes arrive in Australia? Why? Do dingoes still exist elsewhere in the world? Are Australian dingoes expatriate Thai dingoes? Of all the multitudes of animals associated with mankind, why was the dingo the only species with the propensity to be fully domesticated as a dog? Chapter 2 outlines some of the difficulties with solutions to studying dingoes. Do results justify the means? Chapter 3 defines the physical and taxonomic characteristics of pure dingoes and hybrids (crossbreeds with domestic dogs). How does one define a pure dingo and are there distinct types of alpine, desert and tropical dingoes in Australia? Similarly, do dingo subspecies exist in Asia and elsewhere?

Chapters (4–7) describe how dingoes live in Australian habitats, which include the cool mountainous south-east, the hot central deserts, and the tropical northern wetlands. Are dingoes just translocated wolves or have they adapted to Australian environments? Do they display similar behaviours and strategies throughout this range of habitats? Are dingoes loners or pack animals? Territorial or not? Do dingoes migrate? What are the advantages of being dominant? How do dingoes communicate by howling and urinating? What messages are they conveying, and do other dingoes heed them? What do dingoes eat? Do dingoes co-operate when hunting and does this improve success? Part of a team or a jack-of-all-trades — which is the best survival strategy?

Chapter 8 describes the fluctuations in dingo numbers since their arrival in Australia some 4,000 years ago. How many dingoes are there? Does government and pastoralist management control dingoes or merely harvest them, or perhaps even increase numbers? Is the use of poison, trap and gun justified? What is the role of social behaviour and pathogens in the natural regulation of dingoes?

Chapter 9 examines the interactions between dingoes and their prey and the inevitable conflict with pastoralists, sometimes even with conservationists. Are dingoes responsible for the demise of thylacines (Tasmanian tigers) and other native animals on the Australian mainland? Are dingoes wanton killers of calves and sheep? Do they kill for sport or food alone? Do dingoes benefit pastoralists? Is there a trade-off between (a) stock killed and income lost and (b) pests controlled and income saved? Is there an inverse relationship between the numbers of dingoes and the numbers of pests (rabbits, pigs, rodents)? Is the only good dingo a dead one?

Finally, Chapter 10 examines the pessimistic future of expatriate Asian dingoes. Dingoes have won the battles but are losing the war of survival. Why has crossbreeding become a serious problem now rather than earlier? Is the end of pure dingoes really in sight? Does it matter if dingoes are lost? What can be done? What should be done? How would you react if you had to defend your right to have a pet dog? Should your domestic dog be desexed? 'Save the dingo societies' amount to kind hearts but poor application and philosophy. Legislation won't work. It all boils down to whether or not people want dingoes to continue as a subspecies.

The data in this book were hard-won over the past 30 years or so by a small band of dedicated researchers. Alan Newsome spearheaded dingo research in the mid 1960s, first in central Australia and then the south-eastern Highlands; his CSIRO colleagues included Peter Catling, Brian Green, Lindsay Best, Andy Shipway and myself. Dingo research blossomed thereafter with Brian Coman in Victoria, Bob Harden and John Robertshaw in the New England escarpment and Simon Whitehouse in north-west Australia. More recent studies by John McIlroy in Kosciusko National Park, Peter Marsack and Greg Campbell on the Nullarbor, and myself in Kakadu National Park have culminated with Peter Thomson's classic studies in the Fortescue River region of north-west Australia.

No scientist is complete without assistants; in my case, Peter Hanisch, Harry Wakefield, Mick Burt, Tony Hertog, Keith Newgrain, Geoff Bartram, Ian McMillan, Rod Hodder, Peter Brew, Phil Moore, Ross Ellis, Bruce Honeywell, John Lemon, Ros Perry, Ross Cooper, Terry Smith, Gordon Milne, John Randall, Dean Stephens, Peter Sullivan, Mick Gill and Alessia Ortolani to name just a few.

Thanks are due to the pastoralists upon whose properties we often worked; some were initially angry, many remained puzzled, but all were always magnificently hospitable and helpful; in particular Sid Staines (Erldunda), Jim and Ailsa Turner (The Garden), June and the late Rex Lowe (Mt Dare), and the late Mrs Ward (Banka Banka).

The doggers cheerfully shared their knowledge about dingoes and otherwise taught us more than a thing or two about life; Wicky Walsh, Ken Coulson, Syd Ballard and Tom Allen stand out.

In Thailand, I am most grateful to the late Dr Boonsong Lekagul for his guidance and to the dog meat vendors at Tharae, particularly Buanmuang and Vichit, for their help in providing samples.

Research cannot proceed without money and other logistical assistance. In my case, I am grateful to the Victorian Department of Conservation, Forests & Lands (formerly the Vermin & Noxious Weeds Destruction Board), the Australian Meat Research Corporation and the Australian taxpayer for funding; and the Northern Territory Government for the use of many facilities. I also wish to acknowledge the financial assistance of the Australian Meat Research Corporation in the publication and promotion of this book.

Peter Catling, Brian Green, Peter Thomson, Richard Braithwaite, Alan Newsome, Alan Andersen, Peter Daniels, Tony Hertog and Mary Hertog were generous in reading all or parts of the manuscript and provided constructive comments; Tony Hertog also drew many of the figures, Frank Knight provided the drawings of dingoes, and Wendy

Waggitt typed the awkward Appendices.
I feel honoured to have rubbed shoulders with the above-mentioned friends and colleagues, and I sincerely hope that they and others I cannot remember right now consider this book does justice to their efforts.
This book is dedicated to the memory of Geoff Douglas who got me interested in dingoes in the first place and to Harry Frith who let me at them.

The dingo *(Canis lupus dingo)* is a primitive dog that evolved from a wolf *(Canis lupus pallipes/C. l. arabs)* 6,000–10,000 years ago and became widespread throughout southern Asia. Asian seafarers subsequently introduced dingoes into Indonesia, Borneo, the Philippines, New Guinea, Madagascar and other islands including Australia some 3,500–4,000 years ago. Dingoes eventually colonised all the Australian mainland, probably assisted by the Aborigines who had arrived in Australia at least 15 millennia earlier. Some Aboriginal tribes used dingoes to hunt game, especially kangaroos, wallabies and possums. Dingoes contributed to the demise of the thylacine and other native fauna.

The average adult dingo in Australia stands 570 mm at the shoulder, is 1230 mm long from nose to tail-tip and weighs 15 kg; dingoes are smaller in Asia. The coat colour is typically ginger but varies from sandy-yellow to red-ginger and is occasionally black-and-tan, white or black. Most dingoes have white markings on the feet, tail tip and chest, some have black muzzles and all have pricked ears and bushy tails. Pure dingoes are distinct from similar-looking domestic dogs and hybrids because they breed once a year and have skulls with narrower snouts, larger auditory bullae (ear sounding box) and larger canine (holding) and carnassial (cutting) teeth. Coats with a dark dorsal strip or dappling in the white areas indicate hybrid look-alikes. The inheritance of coat colour in dingoes is simple compared to that of domestic dogs, involving only three of the 12 series of genes.

Most female dingoes become sexually mature at 2 years and have only one oestrus period each year, although some do not breed in droughts. Males in arid Australia also have a seasonal breeding cycle of about 6 months; the inability to breed successfully at other times is more probably due to a lack of seminal fluid than to a lack of sperm. Gestation takes about 63 d and litters of 1–10 pups (the average is 5) are whelped during the winter months, usually in an underground den. Pups usually become independent at 3–4 months or, if in a pack, when the next breeding season begins.

Although dingoes are often seen alone, many such individuals belong to socially integrated packs whose members meet every few days or coalesce during the breeding season to mate and rear pups. At such times scent marking and howling is most pronounced.

Dingoes use scent-posts to indicate currently shared hunting-grounds, to mark territorial boundaries, and possibly to synchronise reproduction between pairs.

Vocalisations include three basic howl types: moans, bark-howls and snuffs. Howling is used for long distance communication and has two purposes — attracting pack members and repelling rivals. Dingoes distinguish these purposes by means of howl responses, sight, physical location, and pheromones (chemical messages) to confirm the identity and perhaps the social status of both the initiating and responding howlers. Basic howl types provide information about the location itself, about the howler, and about group size. Overall, howling is mostly used by members of stable territorial packs (or subunits) especially when packs are using or defending essential resources, particularly oestrus and pregnant females, food and water.

In remote areas where dingoes are not disturbed by human control operations, discrete and stable packs of 3–12 dingoes occupy territories throughout the year. Such packs have distinct male and female hierarchies where rank order is largely determined and maintained by aggressive behaviour, especially within male ranks. The dominant pair may be the only successful breeders but other pack members assist in rearing the pups.

The size of a dingo pack's territory varies with prey resources and terrain but is not correlated with pack size. For individuals, home range size also varies with age. The largest recorded territories (45–113 km^2) and home ranges (mean 77 km^2) occur in the Fortescue River area of north-west Australia. Mean home ranges recorded elsewhere are 25–67 km^2 for arid central Australia, 39 km^2 for tropical northern Australia, and 10–21 km^2 for forested regions of eastern Australia. Most dingoes remain in their birth area, but some, especially young males, disperse and the longest recorded distance for a tagged dingo is 250 km over 10 months in central Australia.

Dingoes eat a diverse range of prey types, from insects to buffalo. However, in a particular region they tend to specialise on the commonest available wildlife prey and change their group size and hunting strategy accordingly to maximise hunting success. For example, packs have greater success than solitary dingoes at hunting kangaroos, and vice versa when hunting rabbits. The main prey are magpie geese, rodents and agile wallabies in the Top End (Kakadu National Park); rabbits, rodents, lizards and red kangaroos in central Australia; euros and red kangaroos in the Fortescue River area; rabbits in the Nullarbor region; and wallabies and wombats in eastern Australia. In Asia most dingoes have a commensal (close) relationship with humans and mostly eat rice, vegetables and other table scraps.

Since the early days of European settlement, dingoes have harassed stock, especially sheep and cattle. However, most attacks occur when native prey are scarce (e.g. during droughts or as a result of human disturbance to habitats); cattle also die during drought and dingoes scavenge on their carcasses. Although dingoes often assist humans in keeping down the numbers of rabbits, feral pigs and other pastoral pests, governments and landholders have attempted to control or eradicate dingoes by offering scalp bonuses, by hunting with trap and gun, and by poisoning and fencing. These attempts have been largely unsatisfactory since most control measures merely harvest populations, or even promote increases in dingo numbers by disrupting the social organisation of packs and prompting an increase in breeding rates. Further, the widespread provision of watering points (dams and bores fed by subterranean water) for stock has encouraged dingoes to move beyond the widely scattered natural waters; the provision of abundant non-native food sources (rabbits in good seasons and cattle carrion during drought) has had the same effect.

Today dingoes are under threat of extinction from another source. In the more settled coastal areas of Australia and increasingly so in outback Australia, the barriers between domestic dogs (feral and urban) and dingoes are being rapidly removed so that cross-breeding is common and the pure dingo gene pool is being swamped. Already in the south-eastern highlands about one third of the populations are cross-breeds (hybrids), and unless there is a radical change in people's attitudes the extinction of pure dingoes seems inevitable.

CHAPTER 1

ORIGIN, ANCESTRY AND WORLD DISTRIBUTION

ASIAN ORIGIN AND DUAL EVOLUTIONARY PATHWAYS

There are few animals whose beginning and subsequent history has been as misinterpreted and clothed in myth as that of the dingo. This confusion arose partly because most scientists did not appreciate the dual evolutionary pathways of the dingo, nor that this intricacy was due to tinkering by mankind in the process of canid evolution. It was also partly due to the curious paucity of oral and written records of the unique association between dingoes and early mankind; plentiful records would have helped us understand the evolution of dingoes.

Be that as it may, recent studies by archaeologists, anthropologists, palaeontologists and zoologists provide irrefutable evidence that dingoes are not indigenous to Australia, that dingoes do not share an antiquity with the marsupial megafauna, and that dingoes were not transported to Australia by Aborigines. The rest of this chapter outlines and integrates this evidence to account for the origin of the dingo and its subsequent ancestral development.

Rather than move backwards through time from the present, as is usually done, this chapter traces dingo origins from their beginnings in Asia forwards and along their dual evolutionary pathways, to show how the primitive canine characteristics were maintained along one path but gave rise to evolutionary novelties and monstrosities via founder effects and genetic drift on the other path.

Dingoes began and evolved in Asia. The earliest known dingo-like canine fossils are from Ban Chiang in north-east Thailand (dated at 5,500 years BP) and from north Vietnam

(5,000 years BP). According to skull morphology, these fossils occupy a place between Asian wolves and modern dingoes in Australia and Thailand. There are three major implications. First, these early canines evolved from a wolf species; as Figure 1.1 shows, its prime candidates were the pale-footed (or Indian) wolf *Canis lupus pallipes* and the Arabian wolf *Canis lupus arabs*.

Second, the evolutionary process was domestication (Box 1.1). The Thai site at Ban Chiang is one of the earliest known sites that indicates that people changed their nomadic hunter-gatherer lifestyle to a sedentary and agricultural subsistence. It is widely accepted that this sedentary life allowed commensal relationships to develop between wild animals and people (Box 1.2), which, in this case, was the start of the domestication of wolves into dingoes and other dogs.

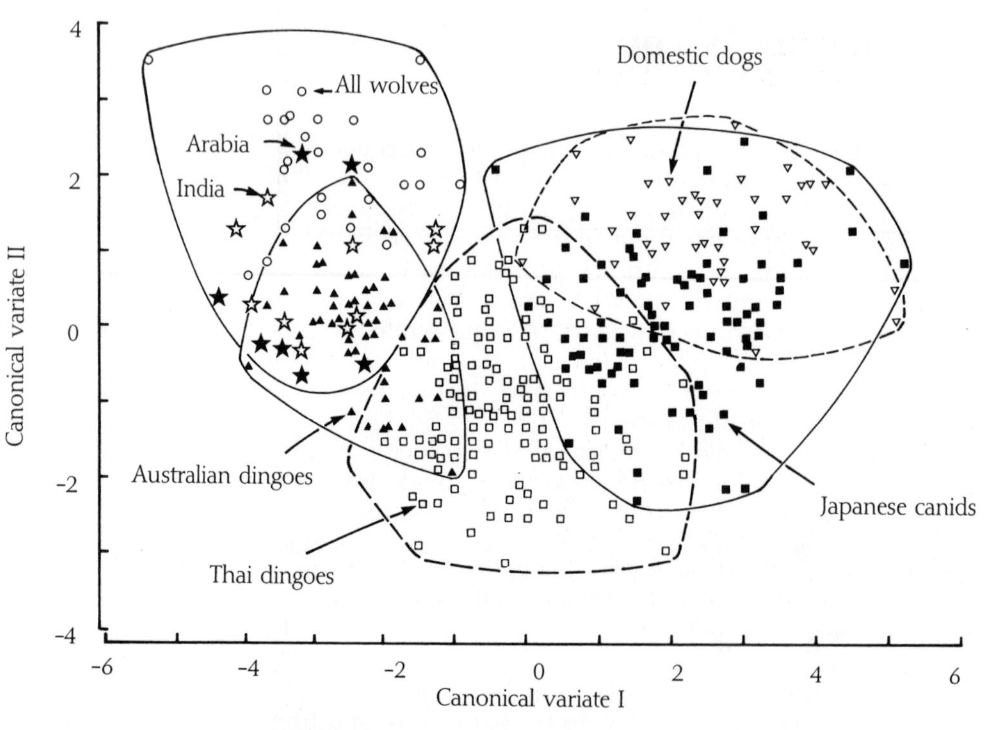

Figure 1.1

Dingo ancestry. Canonical variate analysis of skull measurements that compare wolves (43 skulls from Alaska, Canada, Russia, Sweden, Arabia, Afghanistan, Nepal, Tibet, India, China and Korea), dingoes from central Australia (50), dingoes from Thailand (118), canids from Japan (81 skulls from modern times to 3,000 years BP) and 41 dingo-like domestic dog skulls from Australia. The overlap indicates relatedness, so that of all the wolf species dingoes are most closely related to the Indian wolf (*Canis lupus pallipes*) and the Arabian wolf (*C. l. arabs*).

Box 1.1
The process of domestication

Domestication involves both culture and biology. The cultural process begins when animals are incorporated into the social structure of a human community and become objects of ownership, inheritance, purchase and exchange. The biological process begins when a few 'parent' animals are separated from the wild species and become used to humans. These animals form a founder group which changes over successive generations in response to natural selection under conditions imposed by the human community and its environment, and as a result of artificial selection for economic, cultural or aesthetic reasons.

The morphological changes that occur in domestic animals usually involve a reduction in the size of bones, a reduction in cranial capacity, neoteny and some differentiation in coat colour. Continued selective breeding enhances characters that originally appeared as chance mutations and eventually leads to different breeds.

Domestic breeds are thus analogous to subspecies in the wild in that both are established by reproductive isolation and both are subject to genetic drift; these processes combine to produce genetically unique populations.

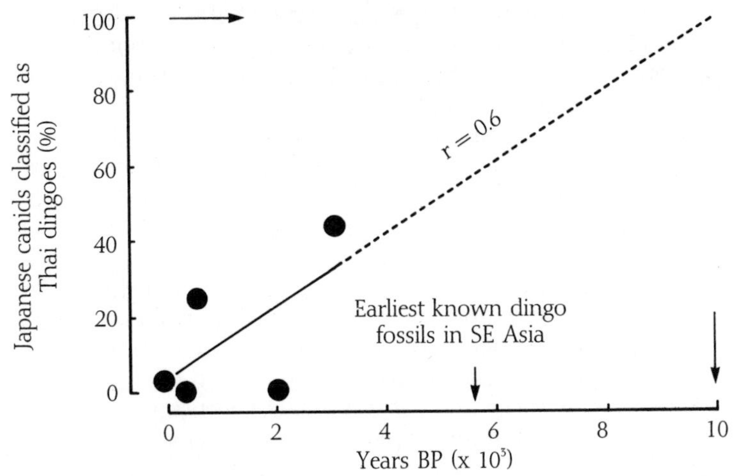

Figure 1.2
Estimating the date of domestication of dingoes. The graph correlates the age of Japanese canids (modern, 200, 600, 2,000 and 3,000 years BP) with their degree of domestication (by comparing them with Thai dingoes, using a principal component analysis of skull measurements). If the graph is extrapolated to the point where 100% of Japanese canids are classified as Thai dingoes (the earliest known dingo fossils, from Thailand and Vietnam, are dated at 5,500 years BP); then dingo domestication began 6,000–10,000 years ago.

Box 1.2
Living together

Symbiosis ('living together') describes a special or close association between individuals of different species, and includes commensalism, mutualism, and parasitism.

Commensalism ('being at the table together') is the process whereby two organisms live in close association but do not depend on each other for survival. However, one partner usually derives benefit; for example, food, protection or transport.

Mutualism is an interspecies relationship that benefits both participating organisms, and is also obligatory to one or both participants.

Parasitism ('eating beside [at the expense of] another') is an association whereby only one participating organism benefits, and at the same time causes harm or annoyance to its 'partner'.

The dingoes in south-east Asia have always shared a commensal relationship with humans, whereas the relationship that dingoes in western Asia and domestic dogs everywhere have with their human owners is a mutual one (mutualism).

Third, the clearcut differences between these early canine fossils and wolves and modern dingoes indicates that domestication was well under way by 5,500 years BP, but probably began no earlier than 10,000 years BP (Figure 1.2).

Somewhere at this stage the evolutionary pathway of dingoes diverged into two distinct streams. During the period of early agricultural settlement between about 9,000 to 5,000 years BP in different parts of the world, it is likely that, for a widely distributed mammal like the wolf, domestication could occur in two ways: many times, in many parts of the world; or from a few sites and by the assisted dispersal over many thousands of kilometres of the initial 'captive' animals and their progeny. In fact, both processes occurred for wolves but separately in western and eastern Asia respectively so that the resultant dual evolutionary pathway characterises the evolution of dingoes and distinguishes it from all others.

Dingo domestication in western Asia

The earliest records of the domestication of wolves are from Israel about 10 millennia ago. These and most other fossils of early domestic 'dogs' (e.g. *Canis familiaris poutiatini, C. f. matris-optimae*) from the Near East and southern Europe have all been described as 'mesolithic dingoes' which suggests that morphologies were similar and that a similar evolutionary process was therefore occurring over a wide geographic area and probably in many human cultures. However, it is impossible to determine whether the initial populations came from local wolves or from partly domesticated wolves that were transported from elsewhere.

What is certain, and important, is that these early primitive canines in western Asia were subjected to intense artificial selection pressures by mankind from the very beginning. Canine-human interactions depicted in early cave paintings, etchings and frescoes in tombs, pyramids and middens suggest that the major reasons for selective breeding were to improve the characteristics of 'dogs' for hunting, herding, hauling, guarding, scavenging and fighting.

Later on, as human societies became more crowded and specialised their labour functions, keeping and creating 'breeds' also involved therapeutic and symbolic values. Thus, for example, a large dog was not only a useful hunting companion but would also ensure its owner an uninterrupted passage through crowded city streets. Similarly, neoteny (the retention of juvenile or ancestral characteristics in the adult animal) set the fashion for city dwellers to acquire novel pets, and although the Pekingese dog, for example, probably developed via deliberate and random selection, its facial features have provided many an owner with a substitute baby.

The ultimate outcome of the many mechanisms of domestication is the immense range of sizes, shapes, colours and temperaments found in modern breeds of dogs from great danes to chihuahuas, pit-bull terriers and bichon frises. What is often forgotten is that this doggie plethora of about 600 true breeding types was derived from a single uniformly structured canine, the dingo, via founder effects, selective breeding and genetic drift.

Dingo evolution in eastern Asia

The evolution of early canines in eastern Asia contrasts starkly with the events in western Asia, described above. Although human societies in the East acquired the early canines for food, hunting, alerting and perhaps other cultural reasons, it seems they were never subjected to selective breeding or other artificial selection pressures. Morphological comparisons between the skulls of the early canines (dated at 5,500 years BP), modern dingoes from Thailand and Australia, and modern dingo-like domestic dogs (Figures 1.1, 1.3) show a great similarity between the dingoes and early canines but a clearcut difference between them and domestic dogs.

The inescapable inference is that dingoes in south-east Asia and Australia have remained virtually unchanged in morphology for at least 5,500 years. This inference is supported by similarities in their body shape, breeding pattern, coat colour range and social behaviours throughout diverse habitats in Australia, Myanmar (Burma), southern China, Laos, Malaysia, various regions in Thailand, Indonesia, Borneo, New Guinea and the Philippines. These data also indicate that dingoes have a wide geographic distribution, at least throughout south-east Asia and Australia.

THE WORLD DISTRIBUTION OF DINGOES AND HOW THEY GOT THERE

The intriguing questions explored in this section are: what is the past and present extent of the distribution of dingoes throughout the world and how did they get there? The answers have much to do with the movements of early humans as their populations expanded and used more of the earth's surface.

Asian seafarers transport dingoes

Over the past 100,000 years or so there has been a succession of human voyages from Asia to other parts of the world

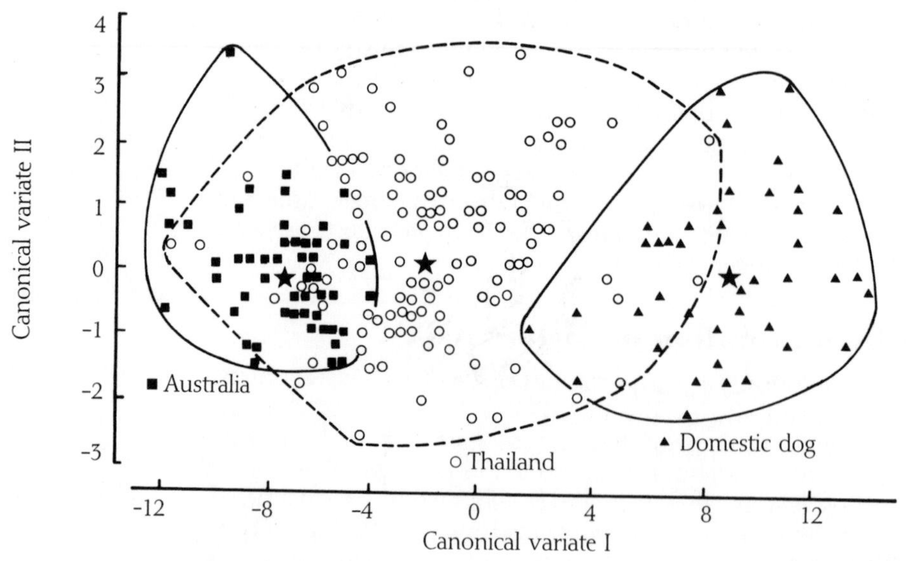

Figure 1.3

Comparing dingoes from Australia and Thailand with domestic dogs by using a canonical variate analysis of skull measurements (adapted from Corbett, 1988). The stars represent the mean values of each group, comprising 50, 118 and 41 skull samples respectively. Although the dogs are breeds closely resembling dingoes in size and conformation, they are distinct.

including Australia (Figure 1.4). The earliest human visitors to Australia were the Pleistocene colonists; the latest were the Macassan trepangers (90–350 years ago), the 'boat people' from Vietnam, and the so-called 'illegal' Indonesian fishermen.

There were probably many visitors in between but none were as crucial to dingo evolution as the Austronesian-speaking people who spread (at a time when dingoes were 'available' for transportation) from mainland Asia through the islands of south-east Asia and into the Pacific. This expansion encompassed the Philippines and eastern Indonesia during 5,000–4,000 years BP, and western Indonesia, western Micronesia and western Polynesia by 3,000 BP; during the first millennium of our era (2,000–1,000 BP) it reached from eastern Polynesia to the island of Madagascar. It was during this expansion that dingoes were transported from mainland Asia and introduced into Australia by Asian seafarers, perhaps on many occasions over many centuries.

Similarly, the movements of seafarers with dingoes supports the conclusion that the primitive 'dogs' of most Pacific

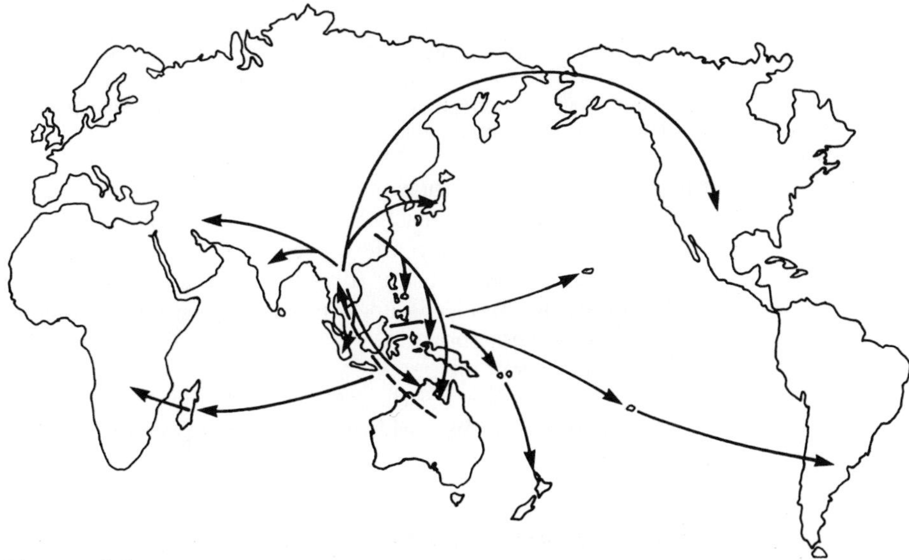

Figure 1.4

Some movements of people with pure dingoes from south-east Asia to Australia and other regions of the world over the past 10,000 years, indicated by the skull morphology of extinct and extant canids and records of human anthropology and archaeology. Most movements were made along sea routes. The distribution of lice on dingoes in Asia and Australia indicates that some dingoes also made the return journey from Australia to Asia (shown by the dashed line).

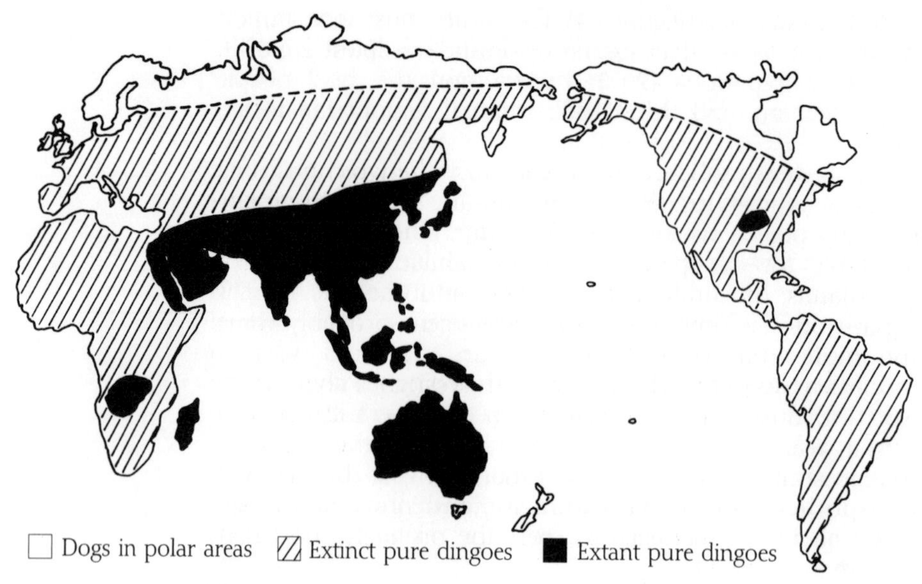

Figure 1.5

The past and present world distribution of dingoes. Pure dingo populations are declining because of increasing numbers of dingo-domestic dog hybrids. In North America, pure dingoes are possibly represented by the 'Carolina dog' which is believed to be a relic of the American Indian dog. The areas showing the probable former distribution of pure dingoes (now extinct), now contain domestic dog breeds that were derived from dingoes. Dogs in the remaining polar areas of the world may have descended directly from northern wolves (*Canis lupus lupus*) or via dingoes from southern wolves (*C. l. pallipes, C. l. arabs*).

islands, including the ancient Kirri dog of New Zealand, were descended from south-east Asian dingoes, while the westward movements explain the presence of the dingo (locally known as the Basenji) in the African Congo.

According to fossil evidence, even the primitive 'dogs' of the Americas were morphologically very similar to dingoes and the movement of people across the Bering Strait could have made it possible for dingoes to reach the Americas; this in turn would give the early dingo a cosmopolitan distribution (Figure 1.5).

Dingo fossils in Australasia

The distribution and antiquity of dingo fossils throughout Asia and Australia fits in with the Asian seafarer theory. The earliest dates in island south-east Asia are at the Niah Cave in Sarawak after 4,500 BP and in Timor between 3,500 and 2,500 BP. In New Guinea, on present evidence, dingoes (inappropriately known as the New Guinea Singing Dog, *Canis hallstromi*) are less than 1,000 years old in the Highlands and no more than 2,000 years old on the south coast. The earliest reliable date for 'dog' in the Pacific islands is about 2,750 BP from the island of Tikopia in the Solomons.

In Australia, the oldest surely dated dingo remains are 3,450 ± 95 BP, and fossils of about this antiquity have been found at sites throughout mainland Australia which suggests that having reached this continent dingoes colonised the mainland quickly and completely. Earlier published claims of dingo remains older than 5,000 years have now been retracted since the stratigraphic integrity of these sites (Devil's Lair, Mt Burr) is doubtful and the specimens probably came from more recent levels.

Another doubtful explanation is that dingoes could have existed in Australia for as long as 11,000 years because Bass Strait was formed about then and therefore prevented them from entering Tasmania. Surely, if the Aborigines brought the dingo in boats to Australia from somewhere overseas, they would also have managed to cross the relatively short distance across Bass Strait.

As indicated above, it was Asians rather than Aborigines who not only transported dingoes to Australia's northern shores but sometimes took dingoes back to Asia. This is suggested by the distribution of a species of biting lice (*Heterodoxus spiniger*) which is found only on dingoes in Asia, but on both dingoes and kangaroos in Australia; the inference is that the lice were transferred to dingoes from kangaroos and spread to the Asian continent on returning dingoes (Figure 1.5). Such back movements of dingoes probably occurred throughout their distribution, effectively mixing the total genetic pot, and thereby also explaining the consistency in dingo morphology over a wide geographic distribution.

Reasons for transporting dingoes

Most probably, dingoes accompanied the Asian seafarers mainly as a source of fresh food during long sea voyages. This hypothesis is based on historic and present relationships between dingoes and people. Dingoes have always been and still are commonly eaten throughout Asia, particularly in the Pacific Islands where dogs are a choice dish, more highly regarded than pig and fowl.

In many areas of rural Asia, particularly in Borneo and northern Thailand, dogs are highly prized as village guard dogs or hunters of game. It seems likely, then, that dingoes were taken on sea voyages in early times for similar reasons, especially since most people probably tended to move along temporary and seasonal coastline campsites. It is not hard to imagine that, on occasions, reproductively mature dingoes wandered from the camps and were subsequently left behind when people moved on. Since dingoes had generalist hunting strategies and an adaptable social organisation, they would be quite capable of surviving and propagating in virtually any 'new' habitats they encountered.

There were probably also cultural reasons for transporting dingoes. Even today in Borneo, dogs and dingoes are killed, eaten and the remains buried with the dead owner as part of a funeral ceremony. In such cultures, death is the commencement of a journey into the afterlife where man is accompanied by his dog; in the same way the dingo accompanied Asian people during their sea voyages.

There are similar accounts of the movements of prehistoric people and their tame wolves or 'dogs' throughout the Americas, from Alaska to Tierra del Fuego, and similar inferences can be drawn.

CULTURAL AND OTHER RELATIONSHIPS BETWEEN DINGOES AND INDIGENOUS PEOPLES IN AUSTRALIA AND ASIA

Aborigines in northern Australia

The Garawa tribe in the Gulf of Carpentaria used dingoes (and later, domestic dogs) to pursue wounded game. As recorded by Pickering, the dingoes would remain quietly with the hunters' extra spears and equipment, which were put down when a target was sighted. When the prey was hit the hunter would call up the dingoes, and they would scent the blood and pursue and harass the animal until the hunter could catch up.

Another technique the Garawa used around replenished swamps and lagoons on the plains in the wet season involved hunters and dingoes driving macropodids (agile wallabies) into boggy or swampy ground where they could be clubbed, or towards the hunters. Sometimes fires were used to assist in the drive; and frilled lizards were often seen in great numbers driven in front of grass fires by flames and smoke; they were caught by clubbing or with dingoes. The Garawa also hunted on moonlit nights for echidnas, rock ringtail possums, northern brushtail possums, northern quolls and black-footed tree rats, and they largely relied on proven hunting dingoes (or dogs). These would have been allowed to run free in the rocky hills. When they had located prey they would make a noise indicating to the hunter that an animal had been caught or bailed up.

Similarly, Thomson recorded the Mildjingi tribe in east Arnhem Land as using dingoes to drive large goannas and native cats into trees so that they could be speared, or dingoes chased bandicoots into hollow logs from which they were cut out with axes. Agile wallabies and antilopine kangaroos were also driven by organised parties of hunters and dingoes so that they could be killed by ambush. Such was the value of the dingo to this tribe that it became one of their sacred totems. Interestingly, this totem has a distinct Indonesian appearance, which almost certainly was derived from association with the Macassar trepangers (and their dingoes) from Sulawesi when they resided in northern Australia each wet season.

According to the explorer Lumholtz, in 1884, the Aborigines in north Queensland reared dingo pups 'with greater care than they bestow on their own children. The dingo is an important member of the family; it sleeps in the huts and gets plenty to eat, not only of meat but also of fruit. Its master never strikes, but merely threatens it. He caresses it like a child, eats the fleas off it, and then kisses it on the snout'.

Aborigines in central Australia

In central Australia (Musgrave, Petermann and George Gill Ranges), in the early 1930s, Finlayson described the Luritja tribe's attitude towards dogs (and presumably dingoes in earlier times) as one of fond indulgence and claimed that their love of dogs and other pets was similar to their love of children. In spite of this regard, however, and unlike the dingoes in Lumholtz's accounts, their dogs apparently lived a most miserable life — they had to hunt all day for their masters, and then most of the night for themselves. Finlayson also wrote that they 'were liable at short notice to be deprived of the doubtful boon of life itself and converted into a scalp, which was traded to a wandering dogger for an issue of flour'. The dingo was thus a utility as well as a pet.

In central Australia, in the 1920–30s (and sometimes in the 1960s) most dingo scalps were obtained by doggers trading with the Aborigines, whose intimate knowledge of the habits of dingoes, and particularly their seasonal movements and breeding places, enabled them to get far better results than a white man alone could. The bonus paid on pups was the same as for adults. The Aborigines were not slow to realise this — they did not molest the breeding females until the time was ripe and obtained a haul of six or eight instead of one, which took the same effort to track and spear. The token of bonus payment was a scalp, and in Luritja country scalps became a sort of currency, much as beavers were used between Indians and the Hudson Bay Company in the Territories of Canada.

In the early 1970s scalps were worth $2 in the Northern Territory and $6 in South Australia, so it is hardly surprising that many a dingo scalp found its way across the border. It so happened that a major study site was based at a cattle station that stretched across the border. The enterprising research team trapped dingoes on the Northern Territory side and sold the scalps to the station lessee for $3. The lessee legitimately claimed $6 from the South Australian Government and then pocketed the $3 difference for little effort. In the meantime, the research team handed in $2 to headquarters as required, but also accumulated the $1 difference for the children's annual Christmas party.

Natives in Papua New Guinea

According to Flamholtz (1986), the Elema tribe believe that one of the gods had a pet dingo named 'Natekari'. This celestial dingo one day caught and ate a cassowary, which he later regurgitated, and from that spot grew taro, which probably is one of the most important food plants in Papua New Guinea. Another tribe, the Orokaiva, believe the dingo originally brought fire to their people. In the Trans-Fly region, the dingo was traded with other tribes. The Orokaiva people, amongst others, regularly used dingoes to hunt wallabies and pigs. The dingoes pursued game and held it at bay until the hunter arrived and it was either speared or slain with a bow and arrow. As a reward, the dingo received a portion of the meat.

Dingoes as food, companions, omens and tutors

However, as in most other parts of Asia, dingoes were regularly eaten. According to some turn-of-the-century explorers, dingoes in Papua New Guinea provided almost all the meat supply. In Cambodia in the late 1970s, dingoes (and dogs) almost disappeared when most of the inhabitants faced starvation under the Khmer Rouge regime. Aboriginal tribes in Australia also ate dingoes. For example, the Garawa tribe in the Gulf of Carpentaria ate dingoes all through the year, but they did not form a staple or reliable part of the diet.

The Akka (Egor) hill-tribe people in northern Thailand believe that a good rice harvest is assured if a lactating female and her pups are eaten in the planting season, even though dingoes seldom breed at that time. Throughout Asia, especially northern Thailand and Indonesia, it is also believed that eating black dingoes will cure fevers and other ailments.

Lumholtz attributed the Australian Aborigines' affection for dingoes to their usefulness in hunting. This explanation also seems to account for the devotion to dingoes exhibited by the Andamanese and Dyaks of north Borneo, and by the seminomadic people of northern Peninsular Malaysia where the dingo is known as the Telomian 'dog'. The Japanese prize the dingoes known as Shiba 'dogs', and believe they accompanied Japan's original settlers (the Jomon people) on their migrations from south-east Asia during the Neolithic period. Another important reason why hunting societies kept dingoes and other wild animals as pets was for their educational value, to teach their children safely about animal behaviour.

Overall, indigenous people respected dingoes and treated them well. Some, such as Aborigines in Australia and Polynesians in Hawaii, even suckled young pups. Some people mourned the death of dingoes, especially those that were good hunters, and sometimes special shrines were erected on the graves of dingoes or their owners. In northern Sulawesi, for example, engravings on the headstones (warugas) of human graves depict the cause of death and the deceased's hobby, character and occupation, and one such headstone, estimated at about 500 years old, depicts a dingo savaging a dwarf buffalo. Apparently the deceased was a hunter who used dingoes to track and bail up the 1 m tall buffalo. In Polynesia, grief over the death of a pet dog was often expressed through tears, poetical eulogies, and ceremonial burial.

Sometimes dingoes were poorly treated. In 12th-century Japan an archery sport called *inuoumono* (dog shooting) was popular among warriors. In this game, one or more dingoes were chased into a circle of mounted archers and shot with bow and arrow according to rules. In the same period, the regent Hojo Takatoki and his followers enjoyed wagering on mass dingo fighting. This was performed by releasing hundreds of dingoes, divided into two groups — the 'reds' and the 'whites' — and letting them fight each other.

A paucity of dingo records

In the five millennia or so that dingoes have been transported from their evolutionary cradle in south-east Asia to virtually all parts of the world, and considering how much impact dingoes and domestic dogs have had on mankind, the cultural and economic associations between dingoes and humans make surprisingly few appearances in the magnificent depictions of human activities and cultures in temples and written records from Pagan to Angkor (but see Plate 1). For example, dingoes do not appear in the huge numbers of frescoes lining the 1,500 or so temples at Pagan in Myanmar except for one series depicting dogs resembling modern borzois that are savaging people. These canids are Kublai Khan's war dogs in action when the Mongols sacked Pagan several centuries ago.

Many records include human associations with other wild animals, many of whom became domesticated for other reasons, but the dingo hardly rates a mention. Even the dingo's ancestor, the Indian wolf, seems more famous in history than the dingo, thanks to the efforts of Mogli, and Romulus and Remus. Of the few available records about dingoes, two, unfortunately, were not recognised as such until very recently.

Dispelling a myth

Aborigines of the Kundi-Djumindju tribe living in the coastal Kimberley region of Australia's Top End sometimes dance a corroboree re-enacting the arrival of dingoes in Australia. They portray dingoes running excitedly up and down the deck of a boat, stopping to look towards the land and down at the water. They jump overboard, paddle ashore, roll in the sand and shake themselves dry, then begin to nose about and hunt.

Most people who have seen this marvellous re-enactment have always presumed that it told the story of how Aboriginal ancestors brought the dingo to Australia. But the story that the dancers tell is about the arrival of the dingo with visitors, not black settlers. These visitors, according to another Aboriginal story, were Baiini — small, yellow-skinned people who made regular visits to Australia and no doubt were the sea gypsies of Indonesia. It is a great pity that the two stories were not linked and accepted as oral history rather than being relegated to Dreamtime myths by European scientists. In doing so, the greatest myth surrounding the origin of the dingo was thus created. Hopefully it is now dispelled forever.

CHAPTER 2

STUDYING DINGOES

Dingoes are not easy to study. They are cunning, highly mobile and elusive, and many operate in rugged terrain, usually at night. Also, the mythical prowess bestowed on the dingo by some pastoralists and doggers has often been unnecessarily daunting to the pioneers of dingo research. This chapter describes some of the techniques used by researchers to unravel some of the dingo's secrets.

CATCHING DINGOES

With a dingo in hand, samples can be taken to determine diet, reproduction, health and other physical attributes; or a live dingo can be released with a radio-transmitter (see Plate 2) that gathers information about its home range, social relationships and other behaviours. But first, one has to catch a dingo.

Traps, snares and cages

In arid regions, leg-hold traps are generally placed at watering points or on the trails leading to them. The positioning of traps is determined by reading dingo tracks and predicting how and from where an animal might approach the water. The trap must be placed downwind of the lure and on the path the dingo will take to investigate the lure. Of the many trap arrangements that have been used, the so-called 'single acute off-set' arrangement (Figure 2.1c) was found to be the most effective by a research team that captured about 2,000 dingoes over 8 years in central Australia.

Modifications include binding the jaws of leg-hold traps with cheesecloth (or non-odorous material) to minimise

Figure 2.1

Three leg-hold trap arrangements: (a) double set, (b) single centre set, and (c) single acute off-set. All traps are buried upwind and close to a trail, track or animal pad which is used by dingoes. Wings (grass tussocks, logs, stones, bushes) are arranged to direct dingoes over the trap. One wing may also be the drag log attached to the trap. A lure (e.g. a mixture of fermenting dingo faeces, urine and blood) is placed upwind of the trap. The distance between the lure and the trap is critical: if it greatly varies from the 'sniff distance' (between the front feet and the nose of a sniffing dingo, about 25–30 cm) for arrangements (a) and (b), slight changes in wind direction become magnified by the time the scent reaches a dingo so that the dingo approaches in a line outside the trap(s). In the single acute off-set arrangement (c) the single wing is almost perpendicular to the wind and angled slightly away from the trail, so the lure must be more than the 'sniff distance' away from the wing. This is usually the most efficient arrangement because it (i) uses only one wing, is more approachable and least likely to arouse wary dingoes; (ii) allows for wind changes in either direction up to 120°; (iii) is less prone to disturbance by large non-target species (e.g. cattle) because they approach on the wind and are large enough to lean over the wing to sniff the lure. The major disadvantage with the other arrangements is their reliance on consistent wind direction to direct a dingo between the wings and over the trap; the double set has a large gap between the two traps and the single centre set has two large gaps on either side of the trap. Other arrangements, such as blind sets (no wings), can be used directly on trails and other areas used solely by dingoes, around carcasses or at scent-posts.

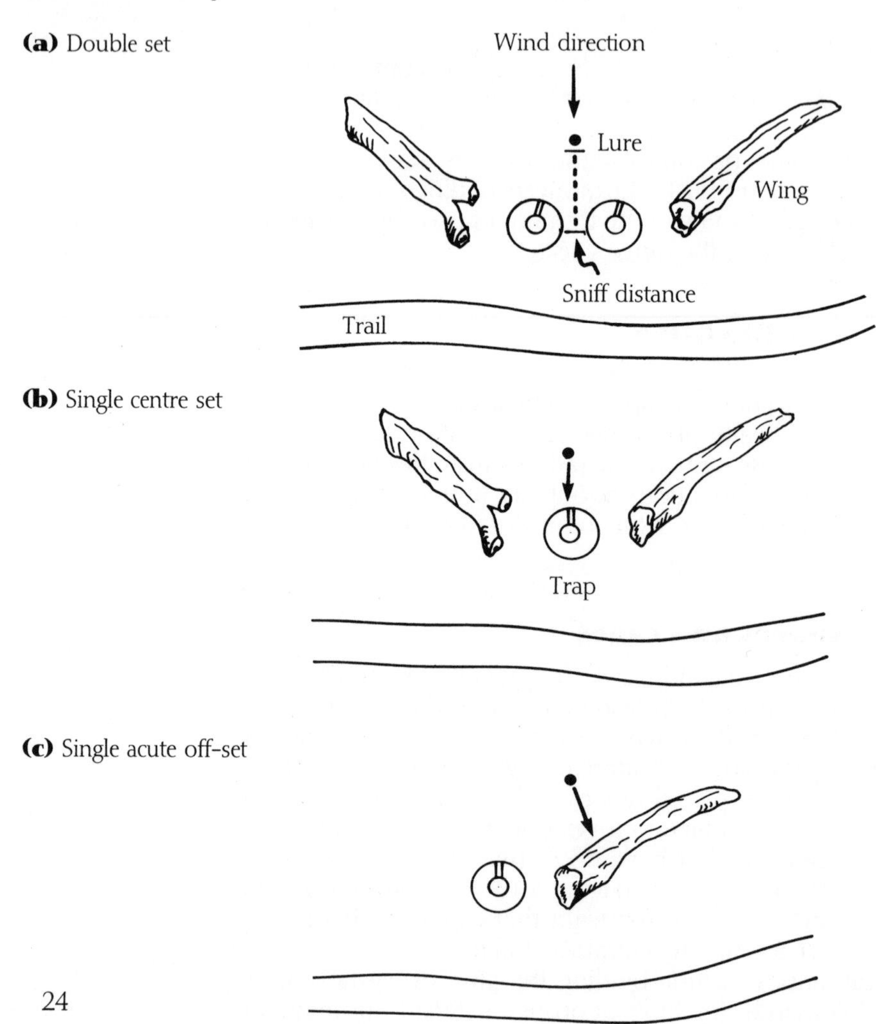

injury to dingoes that are to carry radio-transmitters. An alarm bell system connected between a line of traps and researchers in a nearby camp minimises the time dingoes spend in a trap, and hence minimise injury and stress. Unpadded leg-hold traps are rarely used now because of ethical considerations, and rightly so. Stoppered snares are frequently used instead, especially in forest habitats where dingoes can be channelled to the snare. Cage traps rarely work because adult dingoes are too wary to enter.

Tranquilliser darts

Dingoes can be immobilised with tranquillisers injected via darts shot from guns (percussion, air), blowpipes or long bows. Darting is most effective in relatively open country where dingoes can be pursued by a vehicle or helicopter. In forested country, darted dingoes can sometimes run 100 m or more before the tranquilliser takes effect and become difficult to locate; however, small radio-transmitters in the dart help offset this problem. Darting is probably the best method to capture particular individuals.

Other methods

Capturing pups at natal den sites is relatively easy but the disturbance usually makes the mother shift the pups to other den sites (up to 12 km away) with unknown consequences on subsequent behaviour and research results. Young pups that have left the den can be pursued on foot. Relatively unwary dingoes have been captured in cannon-nets and others have been drugged with tranquillisers in food or water. A rather exciting but inefficient method used by researchers in central Australia was to lasso dingoes with a rope noose on a long pole; the operator was roped to the front of a Toyota. In Thailand, most samples were obtained from 'dog abattoirs' (see Plate 12); some samples were dingoes that had been killed by vehicles, but these were often too decomposed or damaged to use.

OBSERVING DINGOES

Wild dingoes

Dingo behaviour is readily observed from hides atop windmills, trees and other structures, especially in arid areas where dingoes habitually visit scarce watering points. Basic equipment includes binoculars for moonlit nights, infra-red detectors or image intensifiers for dark nights, and a liberal dose of patience. In the heavily forested regions of eastern Australia, focal points include river crossings, clearings, coastal sand dunes and beaches, and other areas where dingoes regularly visit. Such areas are best assessed from preliminary studies of radio-collared animals, so that the observer can anticipate a dingo's movements and take a strategic position for subsequent observations.

At all focal-point areas, 'action' can be initiated or intensified with a lure such as a cattle carcass; however, any unnatural effects this may have on dingo behaviour must be considered.

Captive dingoes

Although there are limits to the value of studying captive behaviour as a model for the behaviour of wild animals, there are advantages such as longer observation periods, accurate and full descriptions, the reliable identification and assessment of individuals' social status, and ready access to biological samples. Taking samples, usually from animals of an identified age, involves regular body measurements, samples of teeth, eye-lenses, bacula (penis bone) and other bones for determining dingo age. It also involves testes biopsies, vaginal smears and semen samples to determine reproductive characteristics; and injecting chemical substances, such as tritium and sodium, to determine water turnover rates in lactation and food requirements respectively. Feeding experiments that determine the amount and composition of the remains of prey in faeces provide correction factors which are used to assess the importance of the different sizes of prey in wild dingo diet.

FAECES AND OTHER INDICATORS

The ability to read bush signs consistently and accurately is an advantage in setting traps and reconstructing events at the scene of a kill. Distinguishing dingo tracks from those of foxes and feral cats is easy, but determining whether they are made by pure dingoes, hybrids or domestic dogs is virtually impossible. However, probably the greatest single aid ever used in studies of wild terrestrial vertebrates is the remarkable range of information gleaned from faeces.

Many animals use faeces to mark territorial boundaries or as a vector for olfactory secretions (pheromones) for chemical communication, thereby providing insight into the structure and maintenance of their social organisation. Since most faeces are species-specific, faecal distribution and accumulation can indicate population size and (for some species) the ratio of males to females. For other species, the weight of individual faecal pellets is linearly related to the weight of the defaecator, so that the age structure of the population and the population biomass can both be deduced. Further, the concentration of progestin in faeces can indicate pregnancy rate, and allow some estimate of population growth rate. The distribution of faeces in different habitats enables one to establish a species' pattern of habitat use, and the analysis of food remains in the faecal matter indicates diet. Finally, the nutritional quality of the diet can sometimes be deduced from a chemical analysis of the faeces residue.

MAJOR STUDY SITES

Most of the data in this book were derived from 11 major study sites, located in Figure 2.2; their major climatic and physical attributes are given in Table 2.1. Other place names that recur in the text are also shown in Figure 2.2.

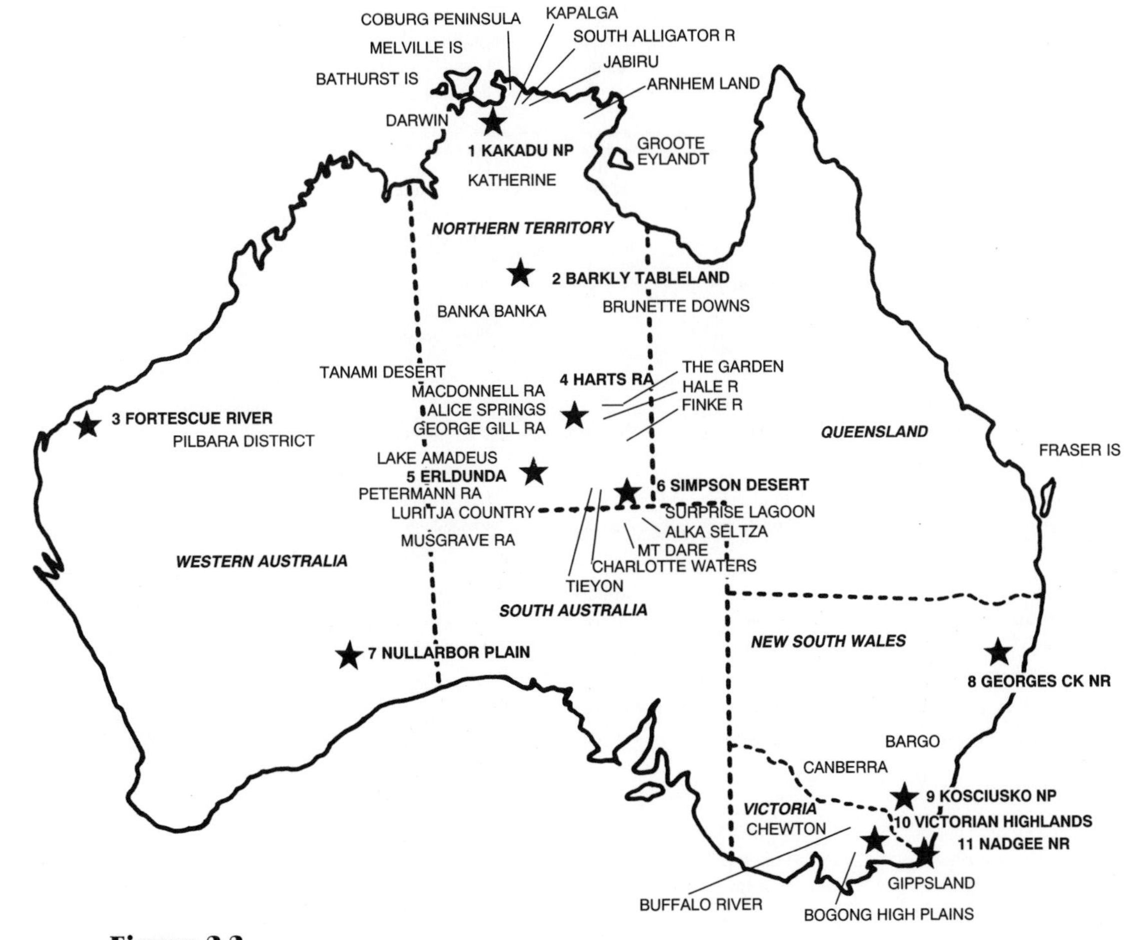

Figure 2.2
Major dingo study sites (★, 1–11) and other locations referred to in the text.

Table 2.1
Climatic and physical attributes of 11 major dingo study sites

	Study Site	Latitude Longitude	General climate	Temp. (°C) mean min-max	Rain (mm) mean annual & other characteristics	Terrain
1	North coastal Australia Kakadu National Park (Kapalga)	12° 37′S 132° 25′E	Seasonally dry monsoonal tropics	18 (Jul) 37 (Oct)	1400 80% Dec-Mar High rel. humidity	Two major habitats (1) seasonally inundated floodplains (2) dry forest & woodlands
2	North-central Australia Barkly Tableland	18° 48′S 134° 02′E	Semi-arid sub- monsoonal tropics	11 (Jul) 38 (Nov-Feb)	875 90% in summer Occasional droughts	Open treeless flat plains with mitchell and flinders grasses on deep black cracking clay soil
3	North-west Australia Fortescue River region	21° 22′S 116° 11′E	Semi-arid tropics	15 (Jul) 35-40 (Oct-Apr)	266-387 Bimodal distribution Jan-Mar, May-Jun	Two major landforms (1) undulating stony plains with dissected hills to 150 m (2) coastal alluvial plains
4	Central Australia Ranges & adjacent plains Harts Ranges Macdonnell Ranges	23° 17′S 134° 25′E	Semi-arid desert	5 (Jul) 35 (Jan) Range -6 to 46	250 Mostly summer rain High evaporation Frequent droughts	Ranges: narrow river & creek valleys surrounded by low hills to steep ranges (700-1100 m) Plains: undulating, sparse to dense woodlands
5	Central Australia Shrublands and steppe Erldunda	25° 13′S 133° 11′E	Semi-arid desert	6 (Jul) 36 (Jan)	173 Mostly summer rain High evaporation Frequent droughts	Sparse mulga shrubland, open chaenopod plains with limestone outcrops, ephemeral saltlakes, red sand dune fields
6	Central Australia Simpson Desert	26° 4′S 135° 15′E	Arid desert	7 (Jul) 40 (Jan)	100 Mostly late summer rain High evaporation Frequent droughts	Two major habitats (1) gibber plateau dissected by small drainage systems (2) large parallel dunes of red sand
7	South-west Australia Nullarbor Plain	30° 30′S 126°10′E	Semi-arid desert	6 (Jul) 34 (Feb)	< 200 Non-seasonal Frequent droughts	Limestone plateau with small shrubs & forbs, some woodland
8	East Australia George's Creek Nature Reserve	30° 37′S 152° 20′E	Cool sub-tropics	-1 (Jul) 32 (Dec)	1200-2500 40% falls in summer	Rugged forested escarpment (150-1400 m) dissected by steep drainage systems
9	South-east Australia Kosciusko National Park	36° 27′S 148° 16′E	Cool temperate	-4 (Jul) 20 (Jan)	750-2000 Mostly winter rain Snow above 1400 m	Montane & subalpine woodlands on mountains (to 2227 m) interspersed with narrow grassy & heathy valleys
10	South-east Australia Victorian Highlands	37° 6′S 147° 36′E	Cool temperate	0 (Jul) 27 (Jan)	650-1940 Mostly winter rain Snow above 900-1200 m for 3-5 months Occasional wildfires	Forested volcanic plateau (up to 2000 m) dissected by steep drainage systems
11	South-east Australia Nadgee Nature Reserve	37° 26′S 149° 55′E	Cool temperate	7 (Jul) 23 (Jan)	750-1000 Mostly winter rain Occasional wildfires	Two major habitats (1) upland (to 540 m) sclerophyll forest & woodland (2) coastal heathland, lakes & swamps

CHAPTER 3

CHARACTERISTICS AND IDENTITY

This chapter describes the basic physical and physiological make-up of dingoes. What makes them distinct from domestic dogs and other related canids? (See Box 3.1.) Differences between regional populations of dingoes in Australia, and between dingoes in Australia and in Asia are also examined to assess whether or not subspecies exist.

Box 3.1
Taxonomic relationships between dingoes and other carnivores and canids

Dingoes belong in the order Carnivora, as a member of the dog family Canidae. This order also contains the families Ursidae (bears), Felidae (cats), Mustelidae (weasels), Procyonidae (raccoons), Viverridae (civets) and Hyaenidae (hyenas).

The family Canidae includes 56 extinct and 13 extant genera. The extant genera, comprising 34 species (223 subspecies) are *Canis* (wolves, jackals and dogs), *Alopex* (Arctic fox), *Urocyon* (grey foxes), *Otocyon* (bat-eared fox), *Fennecus* (fennec fox), *Dusicyon* (zorros and culpeo), *Cerdocyon* (crab-eating zorro), *Vulpes* (red and other foxes), *Cuon* (dhole), *Speothos* (bush dog), *Lycaon* (African hunting dog), *Chrysocyon* (maned wolf), *and Nyctereutes* (raccoon dog).

The genus *Canis* comprises seven species: *Canis adustus* (side-striped jackal), *Canis aureus* (golden jackal), *Canis mesomelas* (black-backed jackal), *Canis simensis* (Simien jackal), *Canis latrans* (coyote), *Canis rufus* (red wolf) and *Canis lupus* (grey wolf).

Within the species *Canis lupus*, 26 subspecies are recognised, including the Australian dingo, *Canis lupus dingo.*

Table 3.1

Linear body measurements of 140 adult dingoes in Australia and Thailand.
Values are means and standard deviations expressed in mm, except weight which is kg

Location:	Victorian Highlands		Central Australia		Kakadu National Park		Thailand	
(Latitude):	(37° S)		(24° S)		(13° s)		(17° N)	
Sex:	Male	Female	Male	Female	Male	Female	Male	Female
Measurement								
Head length	220.4 ± 5.0	214.9 ± 12.0	220.8 ± 4.3	213.6 ± 2.9	226.0 ± 8.0	215.0 ± 4.7	199.7 ± 11.4	184.5 ± 6.0
Ear length	105.8 ± 3.3	101.2 ± 3.2	103.9 ± 3.4	99.5 ± 3.0	101.9 ± 2.9	94.7 ± 4.2	85.0 ± 4.8	78.7 ± 4.0
Shoulder height	580.4 ± 25.9	562.9 ± 25.0	587.4 ± 18.0	559.0 ± 13.0	592.9 ± 23.5	565.0 ± 14.1	516.5 ± 31.6	469.7 ± 23.0
Hindfoot length	191.8 ± 7.8	179.4 ± 6.2	188.0 ± 5.3	179.5 ± 4.6	198.8 ± 10.8	185.0 ± 6.6	164.8 ± 10.8	151.7 ± 6.3
Tail length	299.4 ± 59.1	289.2 ± 44.8	324.3 ± 13.5	311.8 ± 18.1	322.5 ± 24.8	312.9 ± 11.5	262.3 ± 23.6	239.2 ± 17.3
Total length	1244.7 ± 76.2	1218.8 ± 64.1	1207.0 ± 32.4	1168.5 ± 25.5	1288.9 ± 84.4	1271.1 ± 27.7	1078.0 ± 70.4	997.7 ± 39.8
Body weight	15.5 ± 2.0	14.7 ± 1.7	14.5 ± 1.5	12.4 ± 1.1	17.4 ± 1.9	15.2 ± 1.1	12.4 ± 2.4	10.1 ± 1.6
No. of dingoes:	12	16	25	25	12	7	24	19

BODY MEASUREMENTS

The average mature dingo in Australia stands about 570 mm at the shoulder and weighs about 15 kg. Males are larger and heavier than females in all standard body measurements in all regions of Australia and in Thailand (Table 3.1). The largest dingoes have been recorded in (1) Kakadu National Park in northern tropical Australia (maximum shoulder height 645 mm, 22 kg; (2) in the Fortescue River region of north-west Australia (mean weights for males 18.9 kg, for females 15.2 kg); and (3) in the Victorian Highlands where wild canids up to 24 kg have been trapped. The smallest adult dingoes in Australia have been recorded from the central desert regions.

The regional variations in body size throughout Australia may be related to how easily prey is caught. For example, the large size of the Fortescue dingoes may have evolved in response to opportunities to hunt large macropodids.

Thai dingoes of both sexes are significantly smaller than Australian dingoes in all seven body measurements (Table 3.1). This may be a consequence of their diet which is essentially carbohydrate compared to the protein diet of dingoes in Australia.

COAT COLOURS

Evidence from experimental studies (Box 3.3) and wild canids suggests that pure dingoes have only four basic coat colours: ginger, black-and-tan, black or white; all others indicate hybrids and/or domestic dogs (Box 3.2). The distribution of these colours in Australia and south-east Asia and the predominance of the ginger colour is shown in Table 3.2.

Table 3.2
Predominant coat colours (%) of 4573 dingoes (and hybrids) in Australia and south-east Asia.
For Australia, numbers in parentheses refer to study sites shown in Table 2.1

	Ginger	Black-and-Tan	Black	White	Sable	Brindle	Patchy	No. of animals
Australia (study site)								
North (1)	99	0	0	0	<1	0	<1	500
Central (2,4,5,6)	88	5	0	4	2	<1	<1	1320
West (3)	72	15	0	7	0	3	4	256
South-east (10)	43	26	3	<1	14	7	6	734
Total Australia	74	12	1	2	6	3	3	3129
Thailand								
South (Songkhla)	80	5	3	2	1	1	8	216
Central (Bangkok)	63	1	11	5	2	2	16	375
North-east (Sakon Nakon)	57	2	18	7	3	2	10	176
North (Chiengmai)	46	1	16	7	5	3	22	150
Total Thailand	63	2	11	5	2	2	14	917
Indonesia								
Java	81	0	4	0	0	0	15	26
Sulawesi	74	18	0	6	1	2	0	398
Total Indonesia	74	17	<1	5	1	2	1	424
Malaysia (Peninsular)	85	0	0	0	0	0	15	13
Laos	38	1	8	7	8	0	37	71
Myanmar	79	0	5	1	0	0	15	101
Philippines	52	1	10	4	1	1	31	237

In Australia, the frequency of ginger dingoes decreases from north to south, in an approximately inverse relation to human population density and the time that Europeans have occupied the country. As discussed in Chapter 10, this implies that there are more hybrids in south-east Australia.

In Thailand, the trend is the reverse — that is, the frequency of ginger dingoes decreases from south to north, but probably the reason for this is similar. The greater frequency of hybrids in the northern and north-eastern regions is probably due to their proximity to Laos where domestic dogs (and hybrids) are abundant. Gloger's rule — that colours tend to change from dark to light with an increase in latitude — is unlikely to apply to Thai dingoes, since it does not hold true for the southern hemisphere.

In Laos and the Philippines there are many non-dingo colours, especially in patchy combinations. This tendency indicates much hybridisation with domestic dogs and reflects

Box 3.2
Definitions of coat colours

The range of coat colours of dingoes (D) and their hybrids (H) with domestic dogs can usually be assigned to one of the following categories:

Ginger (D,H). All shades from deep red to light sandy ginger, sometimes cream. Uniform dorsal coloration, ventrum usually lighter. Most have white toes, feet or socks and a white tail tip. Some have dark muzzles. See Plates 3 and 4.

Black-and-tan (D,H). Black with tan areas on cheeks, ears and legs. Some have white chests and/or white toes or feet, and/or a white tail tip. See Plate 5.

Black (D,H). Usually a uniform coloration. Some may have a small or large white chest patch.

White (D,H). Uniform coloration (not albino). See Plate 6.

Sable (H). Black or dark dorsum extending to about halfway down the sides, ginger ventrum. Some have white chest patches, and/or white toes or feet, and/or white tail tips. See Plate 7.

Brindle (H). Dark narrow vertical stripes, often broken, on the dorsum and sides of the body and legs, sometimes including the face. Some have white chest patches, and/or white toes or feet, and/or white tail tips. See Plate 8.

Patchy (H). Either ginger or black with large and/or small white patches on dorsal, ventral and leg areas. Some have a white tail tip. See Plate 10.

the long domination of these countries by European and American colonists, respectively, and their domestic dogs (see also Chapter 10).

Black dingoes

The worldwide distribution of black coloration is interesting, but puzzling. Solid black dingoes, widespread throughout Asia, are often the second most frequent coloration, but only rarely recorded in Australia in recent times (Table 3.2). This coloration may once have been more widespread. One of the first descriptions of dingoes in Australia, by Collins in 1798, described one from the Sydney area that was 'quite black'. In 1832 the explorer Mitchell saw 'a small black native dog' in northern central New South Wales; and in 1889 another explorer, Lumholtz, noted that the black variety (with white chest) generally appeared in Queensland. The Horn scientific expedition to central Australia in 1894 recorded that 'most of the (dingo) specimens were of the yellow-brown colour, but occasionally they were black'. Dingoes with large proportions of black coats were also recorded around the early

Box 3.3
The inheritance of coat colour in domestic dogs and dingoes

The inheritance of colour by domestic dogs is complex and imperfectly understood, probably because of their polyphyletic origins (ie several ancestral species) and intense artificial selection pressure by humans since Mesolithic times. However, most geneticists agree that about 12 series of alleles are involved. The number of alleles in each series varies, but altogether there are over 30 genes whose interplay explains the array of coloration seen in domestic dogs. The generalised (hypothetical) ancestor of the dog, however, is assumed to have just 9 series operating, and its genotype is indicated in the table below. All of the series are homozygous (unvarying) and the hypothetical ancestral wild dog would have been a uniform black. An asterisk indicates the series that have been detected in dingoes.

Series	*Genotype*	*Phenotype*
A*	a^wa^w	Agouti
B	BB	Black
C*	CC	Full colour
D	DD	Intense
E	EE	Extension of black
G*	gg	Absence of greying
M	mm	Non-merle (dappling)
S*	SS	No spotting
T	tt	No ticking

In determining the inheritance of dingo coat colours, researchers examined differences from the basic genotype by experimenting in crossbreeding and backcrossing dingoes and domestic dogs in captivity. These experiments indicated that most variation occurs at only three of the loci (A, C, S series).

The *A series* controls the production of ginger and black pigments. In dingoes, the ginger and black-and-tan colour patterns breed true, but ginger is dominant to black-and-tan in a simple 3:1 ratio. Backcrosses of these F_1 to animals with the homozygous recessive coloration results in ginger and black-and-tan progeny in a 1:1 ratio. There are no data available for solid black dingoes. This relatively rare coloration may be due to either the masking of most genes in the agouti (A) series or a fourth series (B).

The *C series* is concerned with the strength or dilution of full colour. Hence, in dingoes, there are varying shades of ginger ranging from red-ginger to creamy and white.

continued ...

Although creamy and white dingoes breed true, both are recessive to ginger.

Allelles within the *S series* determine the presence or absence of white points. Most dingoes have white toes, feet or socks, and tail tips. Other series may be present in uncommon colour variants; for example, ginger dingoes with dark muzzles or ginger-coated dingoes with reddish overtones may be due to the E series.

Crossbreeding experiments in captivity with dingo-like domestic dogs, such as heeler, kelpie and collie, indicate that the F_1 hybrids are very akin to dingoes in both shape and colour. Such hybrids, therefore, can be difficult to recognise in the wild (see Plate 9) although ticking or spotting in white areas is usually prominent in dingo/heeler hybrids. On the other hand, the F_2 generation and backcrosses to the domestic parent show up as obvious mongrels. Similarly, F_1 hybrids resulting from non-dingo-like domestic dogs, such as labrador and doberman, are relatively easy to identify as mongrels, usually from the shape of the head, ears and tail.

In summary, crossbreeding experiments in captivity in conjunction with data on reproduction and skull morphometrics indicate that the basic range of coat colours of pure dingoes is ginger, black-and-tan, black or white (see Plates 3, 5, 6). Hybrids range from patchy (ginger-and-white, black-and-white), sable, brindle and other colours (see Plates 10, 7, 8).

settlements of Western Australia in 1842. In 1863 the famous naturalist, John Gould, commented that black was one of the normal dingo colours. That black specimens are pure dingoes is indicated by the similarity of skulls of black specimens from Thailand to those of pure ginger-coloured dingoes.

Black-and-tan dingoes

Despite the fact that skull morphology confirms their identity, there is still controversy over whether or not black-and-tan represents a pure dingo coloration. Today, black-and-tans occur most frequently in Australia (especially southern Australia) and Sulawesi. But it is remarkable that there are no references to 'tan' colouring on the black or partly black animals recorded by the early explorers and settlers in Australia in the last century; perhaps this was an oversight on their part.

Another explanation is that the genetic expression of black-and-tan dingoes was masked when they bred with ginger dingoes (Box 3.3), so that they were rarely encountered. This notion is supported by recent studies of relatively undisturbed and thus stable populations in which small pockets of black-and-tan (and white) dingoes exist in a 'sea' of ginger dingoes. For example, in the Fortescue River region of north-west Australia the close genetic relationships between pack members and the stability of packs is indicated by the frequencies of the coat colours of five packs living in adjacent territories. In a pack of 33, 21.2% were black-and-tan and 48.5% were white, compared with the 20% black-and-tan and 2.5% white of the 80 dingoes comprising the four other packs.

REPRODUCTION

Breeding pattern

Dingoes produce only one litter of pups each year, and except for tropical habitats, litters are whelped at about the same time each year (Table 3.5). This characteristic breeding pattern is determined by the fact that most females have an annual oestrus cycle, whereas males can breed continuously in all habitats except hot and arid regions where they have an annual testis cycle. The precise onset and extent of breeding, however, varies with age, social status, geographic latitude, seasonal conditions, and whether animals are pure dingoes or hybrid.

Female dingoes

In central Australia most wild females commence breeding when they are 2 years old and usually mate in April/May, about one month earlier than most of their counterparts in southern temperate Australia. In stable packs the most dominant (alpha) female, usually the oldest, tends to come into oestrus before other females in the pack, and some subordinate females seem to go through a pseudopregnancy. During droughts in central Australia, all young (<1 y) females and some older females do not breed at all, and for those adults that do, the onset of breeding is delayed by about 2 months (Figure 3.1).

Male dingoes

Males reach full sexual maturity at 1–3 years of age. In hot and arid central Australia, males appear to be true seasonal breeders with maximum testis weights occurring about April, the peak of the mating season (Figure 3.2b). During the non-breeding season, captive males do not show any sexual interest in oestrus domestic females but readily mate and sire pups with the same domestic females during the dingo breeding season.

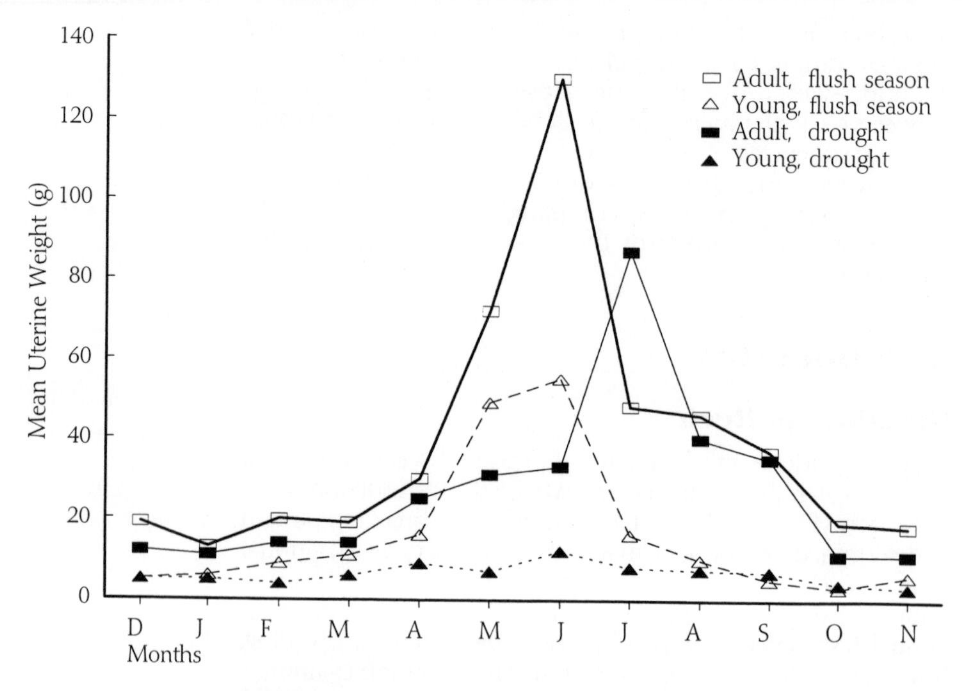

Figure 3.1

The breeding cycle of adult (>1 y) and young (<1 y) female dingoes in central Australia (adapted from Catling et al., 1992). In flush seasons most adult females are heavily pregnant by May-July, but not all young females become pregnant, as indicated by their low uterine weights. In droughts young females do not breed at all, and breeding is delayed by about 2 months for the older females that do breed.

In contrast, males in the cooler temperate highlands of south-east Australia can breed throughout the year, and many successfully sire pups with domestic bitches in the dingo non-breeding season. One explanation for this difference is outlined below.

Studies of captive males and analyses of testis sections of wild adult males (>1 y) in central Australia indicate that spermatogenesis occurs throughout the year but the ejaculate has significant numbers of sperm only for part of the year — from February to July (Figure 3.2c, d). Since there is a sharp peak in the monthly weights of the prostate gland compared to those of the testes (Figure 3.2a, b), the relative lack of seminal fluid (the bulk of semen) from the prostate gland probably accounts for the cyclicity of the breeding pattern of males in central Australia. If this is related to environmental factors, particularly temperature, then this explains why male dingoes from this region become capable of breeding continuously when transported to Canberra in temperate south-east Australia.

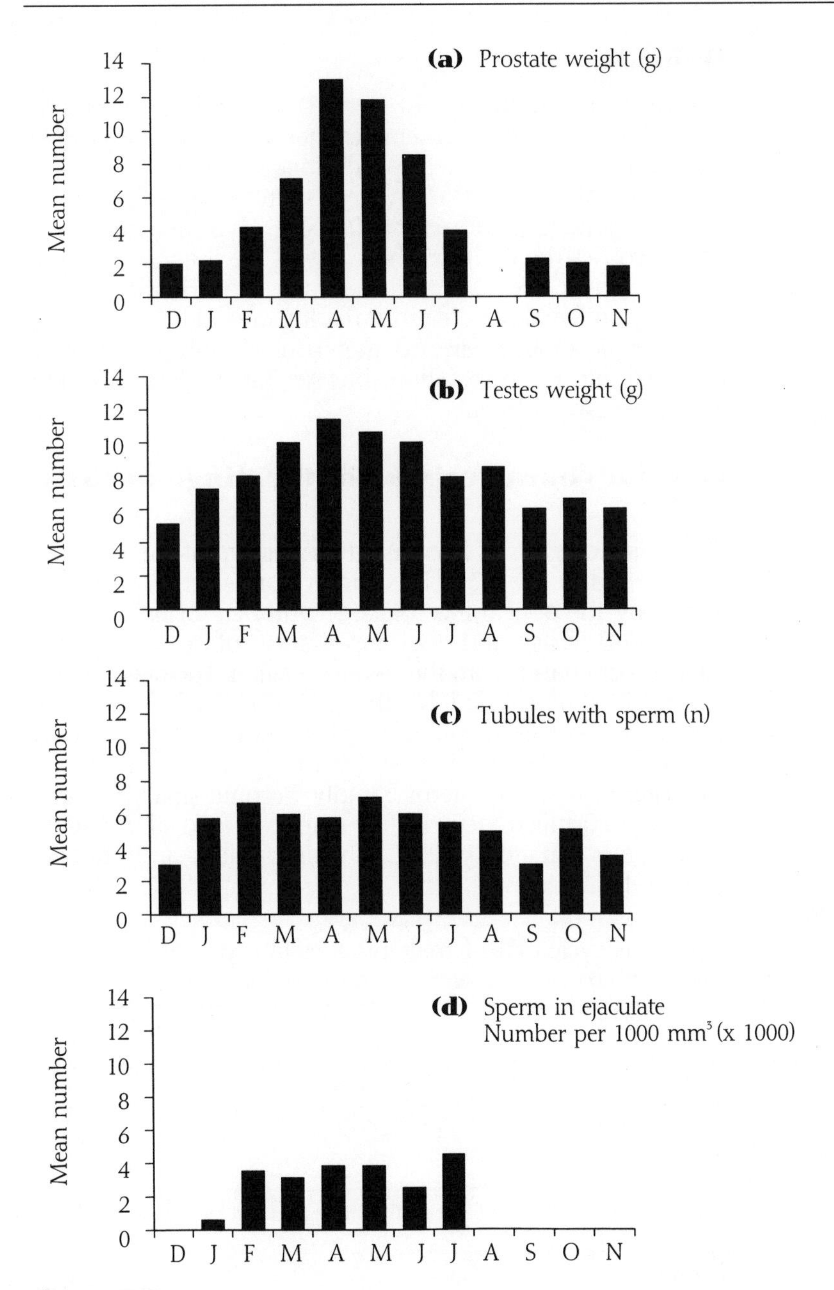

Figure 3.2

Cyclicity in reproductive parameters for adult (>1 y) male dingoes in central Australia (adapted from Catling et al., 1992): (a) weight of prostate gland; its secretion is a vital component of semen and may help 'activate' sperm; (b) weight of testes, where sperm are produced; (c) presence of sperm in testes, indicating spermatogenesis; (d) number of 'activated' sperm in the ejaculate, indicating breeding potency. Only very small amounts of spermless ejaculate were extracted in Aug-Dec. Data were derived from wild (a, b, c) and captive (d) dingoes.

Hybrids

The reproductive pattern of hybrids (F_1 to third generation backcrosses to dingoes) resembles that of most domestic dogs and is unlike that of most pure dingoes. Hybrid males can breed all year, unlike male dingoes in central Australia but like male dingoes in south-east Australia. Hybrid females can have two oestrus cycles each year and are thus theoretically capable of producing two litters each year. However, most wild hybrid females are flat out rearing one litter each year, as pure female dingoes are, so that usually only one litter is recruited into the population (but see Table 3.5 for breeding in south-east Australia).

Cellular characteristics of the dingo oestrus cycle

The major diagnostic features of each stage of the cycle are summarised in Table 3.3. The onset of pro-oestrus is characterised by the appearance of large numbers of erythrocytes (red blood cells) and the appearance of a sanguineous (bloody) discharge from the swollen vagina. The beginning of oestrus is characterised by the prevalence of many large, angular and cornified epithelial cells and the disappearance of leucocytes (white blood cells). Oestrus is finished when the epithelial cells of the uterus rapidly become small, rounded and non-cornified, leucocytes reappear, and the vaginal discharge is dark or yellow, sometimes clear, and contains much cellular debris.

Table 3.3

Characteristics of the oestrus cycle of 10 female dingoes in captivity in Alice Springs as determined by microscopic examination of vaginal smears

Stage of Cycle	Cell types			External signs		Duration (days)	
	Epithelial	Leucocytes	Erythrocytes	Vulval tumescence	Vaginal discharge	Mean	Range
Anoestrus	All non-cornified, small, rounded	Present	Absent	None	None	–	–
Pro-oestrus	Most non-cornified, becoming large, angular and cornified	Present in some females	Present in excessive numbers	Maximum size	Sanguineous	11	10-12
Oestrus	Most large, angular, cornified	Absent	Present in some females	Maximum size	Sanguineous, variable composition	10	7-16
Met-oestrus	Most non-cornified, becoming small and rounded	Present	Present	Variable decrease	Variable volume and composition	–	–

In Alice Springs pro-oestrus for female dingoes in captivity lasted about 11 d, and oestrus lasted about 10 d; and both values are similar to the values for domestic dogs. It is interesting that two captive females that were kept by themselves (with reproductively active dingoes in adjacent pens) remained in pro-oestrus for over 2 months. There are no equivalent data for wild females in central Australia, but behavioural data indicates a likelihood of similar values — about 10–12 d each for pro-oestrus and oestrus.

However, in the Fortescue River region of north-west Australia, behavioural data suggested that pro-oestrus may have lasted about 30–60 d, which is consistent with data for wolves (mean 42 d) and coyotes (60–90 d). It has been suggested that a protracted pro-oestrus period is necessary for wolves and coyotes to form and/or strengthen pair bonds, or to induce full sexual potency in males. This may also have been the case for the Fortescue River females. However, another explanation — based on the two 'isolated' captive females in Alice Springs — is that the Fortescue River females were of low social rank and thus denied reproductive access to a male (see also Chapter 6).

Gestation, litters and sex ratio

Gestation, the time between conception and birth, lasts 61–69 days for captive dingoes, which tends to be several days longer than for most hybrids and domestic dogs. The average litter size is about five throughout different habitats in Australia and Thailand, and usually there are slightly more males born than females (Table 3.4). The largest litters in canids (average 7–10) are found in the African hunting dog.

Most dingo litters are born in the dry, cool or winter months in both the northern (Oct-Dec) and southern (June-Aug) hemispheres (Table 3.5). At the equator (Sulawesi), it seems that dingoes may breed in all months, and although more data are required for confirmation, this notion is supported by data from sites relatively close to the equator (Thailand and northern Australia) where litters of pure dingoes have been recorded in most months. Also, as indicated above, dingoes in Australia mate slightly earlier in arid regions than they do in more southern temperate regions. The larger range of months in which litters are born in south-east Australia, particularly the summer and autumn months (Nov-Apr) is almost certainly due to the presence of hybrid females.

African hunting dogs have a similar breeding pattern to dingoes. In Kenya, on the equator, pups are born in every month except September; however, in Kruger National Park in South Africa (24°S), breeding is restricted to the dry season.

Similarly, dholes in sub-tropical India are not confined to a narrow breeding season and pups are born between September and January.

Table 3.4
Gestation (days), litter size and sex ratio

	Gestation		Litter size		Sex ratio
	Mean	Range	Mean	Range	(♂:♀)
Wild dingoes: Australia					
Northern Australia	–	–	5.0	4–10	–
Central Australia	–	–	5.0	1–9	1:0.9
Fortescue River region	–	–	5.2	4–7	–
Victorian Highlands	–	–	5.5	2–9	1:0.9
Wild dingoes: Thailand					
North	–	–	4.0	3–7	–
Central	–	–	5.2	1–9	–
South	–	–	5.0	3–7	–
Captive dingoes					
Alice Springs	64	61–69	4.7	1–7	1:0.9
Canberra	–	–	4.0	–	1:1.3
Captive hybrids	64	59–65	4.3	3–5	1:1.0
Domestic dogs					
Alice Springs	63	58–64	5.5	1–8	1:0.6

Table 3.5

Months when dingo litters born (%) in Asia and Australia.
✔ = litters inferred from data; ?✔ = litters inferred from personal communication; - = no or incomplete data

		O	N	D	J	F	M	A	M	J	J	A	S	No. of litters
							%							
Asia	*(latitude)*													
Bangladesh	(25°N)	✔	✔	✔	✔	0	0	0	0	0	0	0	✔	-
Myanmar	(21°N)	✔	✔	✔	✔	0	-	-	-	-	-	-	-	-
Laos	(18°N)	✔	✔	✔	-	-	-	-	-	-	✔	✔	✔	-
Thailand	(14°N)	47	21	2	4	2	9	0	0	0	6	0	9	47
Philippines	(10°N)	✔	✔	✔	✔	✔	0	0	0	-	-	-	-	-
Sulawesi	(0°)	✔	✔	✔	✔	✔	?✔	?✔	?✔	?✔	?✔	?✔	?✔	-
Australia														
North	(13°S)	0	0	4	8	4	0	0	4	29	33	13	4	24
West	(21°S)	0	0	0	0	0	0	0	2	24	39	35	0	46
Central	(24°S)	0	0	0	0	0	0	0	18	26	37	17	2	76
South-east	(37°S)	0	1	1	1	1	3	6	18	27	24	16	3	80

Characteristics of natal dens

In central Australia most dens were in enlarged rabbit-holes, usually with rabbits living in adjacent holes in the same warren (see Plate 13). Other den sites included caves in rocky hills, under debris in dry creek beds, under large tussocks of spinifex, among protruding tree roots and under rock ledges along watercourses. In the Fortescue River region of north-west Australia, where rabbits are absent, most dens were in cave complexes in hilly terrain (some dens had multiple entrances), others were in hollow logs, under spinifex or in enlarged goanna holes. Similarly, in the wet-dry tropics of northern Australia, most dens were found in enlarged burrows of large goannas, with others under fallen trees and other debris. In the moist eastern highlands, most dens are reported to be in hollow logs, old wombat burrows and occasionally in caves under rock ledges.

A common feature of all dens is that most occur very close to water sources, including small perennial springs and soaks. Sometimes several dens occur close to one watering point. For example, each of two watering points in the Harts Ranges in central Australia (about 10 km apart) had three dens spaced about 2.5 km apart.

Doggers and stockmen in central Australia and the Barkly Tablelands often speak of dens being used year after year, particularly cave dens. However, data on known sites and tagged dingoes indicates that the dens themselves are used once only, but the same general area is frequently used — sometimes by the same female or one of her offspring, but usually an unrelated female.

In central Australia the entrances to most dens face north, presumably to receive maximum sunlight in winter and spring. However, in the Fortescue River region significantly more den entrances face away from the north; suggesting that dingoes there display directional preferences for certain den sites, but for unknown reasons. Den sites in arid Australia usually have several deep gutters nearby where the pups play or quickly hide if danger threatens, and there are usually large shady trees nearby which large pups rest under during the hottest part of the day.

Comparison of reproduction in Australian dingoes and wolves

Both species are seasonal breeders and usually mate from January to early April: late winter/early spring for wolves and late summer/early autumn for dingoes. Gestation is similar (wolves: 62 ± 4 d; dingoes: 64 ± 3 d), and so are litter sizes (wolves: average 6, range 1–11; dingoes: average 5, range 1–10). In the wild, female wolves do not breed before they are 22 months old; this is also generally true for captive females, but some have bred or shown reproductive activity at 9–10 months. Wild dingo females also usually breed in their second year but, unlike wolves, most captive females breed during their first year of life, at 9–10 months. In the Alaskan wilds, multiparous wolves breed earlier than first breeders, and a study of captive wolves found that the oldest female came on heat first, followed by another adult 2 weeks later, and then a 6-month old female. Captive dingoes also show this sequence of the oldest females coming on heat before younger females.

In both wolf and dingo packs, wild and captive, usually only one litter per pack is successfully raised, usually that of the most dominant (alpha) pair, indicating the reproductive suppression of other females in the pack. However, the methods of the two species' reproductive suppression are different. Alpha male and female wolves suppress copulation in subordinate pack members so that only the alpha female becomes pregnant, whereas dingoes' main method of suppressing reproduction is dominant female infanticide: all the subordinate females become pregnant and whelp pups, but all are killed by the alpha female (see Chapter 6).

There are also similarities in the composition and hierarchical structure of packs, and in the nature and frequency of aggressive and other interactions between pack members. Such similarities in the social organisation and dynamics of wolf and dingo packs reinforce the conclusion that dingoes have evolved from wolves. However, the differences, such as

Box 3.4
DNA fingerprinting: identifying dingoes by their genes

Genes are segments of the deoxyribonucleic acid (DNA) molecules that make up the chromosomes occurring in the cells of all living organisms. Genes are essentially a set of instructions (chemical messages) that determine the physical appearance and biochemistry of every living organism. The instructions in genes are passed on from one generation to the next, so that offspring inherit a range of individual traits from each of their parents. Thus, every individual, save identical twins, has a unique genetic pattern in their DNA. This is the basis by which molecular biologists can clearly distinguish between closely related varieties (e.g. cabernet sauvignon, sultana, pinot noir and chardonnay grape varieties), and identify individuals (e.g. revealing a child's real father in a paternity suit).

How are DNA fingerprints made? First, DNA is extracted from body tissues or fluids (e.g. blood) and then cut at specific points by using so-called restriction enzymes. The fragments of DNA (called microsatellites), placed on an agarose gel, are separated by running an electric current through the gel, a process known as electrophoresis. The microsatellites have a negative charge, so when a positively charged electrode is placed at the other end of the gel, the charged microsatellites travel towards it through the gel; the shorter, lighter fragments move more quickly through the gel than the longer, heavier fragments do. Thus the microsatellites are separated according to size.

The pattern on the gel, invisible at this stage, is covered by a nylon membrane and a layer of paper towels. The towels draw the microsatellites upwards into the nylon membrane by capillary action. Radioactive DNA probes are then applied to the membrane, and these bind to any complementary microsatellite sequences.

Finally, to make the pattern of bound microsatellite fragments visible, scientists put an x-ray film next to the membrane. The places where the radioactive DNA probe has bound to the microsatellites causes a visible fog on the film. This creates a pattern of bands that look a bit like the bar codes on retail goods. These bands, representing the unique DNA profile of an individual, are commonly called a DNA fingerprint.

For dingoes, DNA fingerprints could be used to determine whether particular individuals are pure dingoes or hybrid; and in the case of hybrids, which parent is hybrid. Similarly, significant differences in dingo populations between and within countries (subspeciation) could be confirmed. DNA fingerprinting could and should also be a vital component of registering dingoes by the Australian National Kennel Council and other Dingo Preservation Societies (see Chapter 10).

dominant female infanticide, are probably a consequence of the capricious Australian environment where drought has shaped a distinct, unique dingo ecology in Australia, explained in following chapters.

BLOOD PROTEINS AND DNA FINGERPRINTING

Although blood proteins (enzymes) — for example, Glucose-6-phosphate-dehydrogenase — were examined at 30 loci, none differed consistently between dingoes and domestic dogs. Future studies involving DNA fingerprinting may prove useful (Box 3.4). Meantime, the most reliable method to identify pure dingoes involves measurements of particular bones and teeth in the upper skull (see Chapters 1 and 10 and Appendix E).

LONGEVITY

Dingoes from central Australia live for up to 10 years in the wild and one male lived for 13 years in captivity. This latter animal was almost blind and almost certainly would not have survived to this age in the wild. In the northern tropics, one adult male, about 4 years old when captured, was recaptured 5 years later. In the Victorian highlands animals up to 12 years old have been trapped but it is unknown whether they were pure dingoes or hybrids.

Methods to estimate the age of dingoes using headlength and eye-lens weight are given in Appendix B. Other methods include the weight or length of bacula (juveniles, immature and mature adult age-classes), the eruption pattern of permanent teeth (useful up to 6 months), tooth wear (6–12 months), the annular cementum bands in the root tissue of teeth (12+ months). Alternative methods use closure of the foramen at the root-tip of canine teeth to distinguish juveniles from older animals; and for dingoes with closed root-tips on all canines, the width of the pulp cavity of canine teeth distinguishes yearlings from adults.

WATER NEEDS

In arid areas dingoes do not need drinking water every day if other sources are available. If food is greatly abundant and catchable, dingoes apparently can absorb sufficient free and metabolic water from prey and live without drinking water, at least during the cool winter months. For example, dingoes have lived in the middle of the Simpson Desert, about 130 km from the nearest water (Surprise Lagoon). Long-haired rats were abundant at the time and studies indicated that they supplied adequate calories and dietary water for dingoes' needs. When the rat populations declined, dingoes left the desert and returned (presumably) to their former haunts to impose on the watering points and food resources of dingo packs living on the edge of the desert, as described in Chapter 6.

In central Australia weaned pups living at den sites several kilometres from drinking water also obtain (absorb) most of their water from food. However, as the weather grows hotter, some females have been observed to carry water in their bellies and regurgitate it to the pups (also described in Chapter 6).

In most mammalian mothers, lactation, not unexpectedly, increases their rate of water turnover, although captive lactating females have almost no such increase because they frequently lick the anal and urethral orifices of small pups while they are suckling, and this stimulates them to void. The female then licks up all the faeces and urine, thus recycling the water they contain as well as keeping the den clean. Wild dingoes probably have higher rates of water turnover because they need to catch their own food.

TAXONOMY: DO SUBSPECIES EXIST?

The naming of dingoes

The first printed use of the word 'dingo' was by Tench in 1789, who stated: ' . . . the only domestic animal they [Aborigines] have is the dog, which in their language is called Dingo, and a good deal resembles the fox of England'. The first pictorial representation appeared in Phillip's *Voyage to Botany Bay*, 1789, entitled the 'dog of New South Wales', and Kerr reproduced this figure in *Animal Kingdom* in 1792 with the name *Canis antarcticus*.

After this, the scientific name of the dingo underwent much synonymy as *Canis dingo* Meyer 1793, *Canis familiaris australasiae* Desmarest 1820, *Canis australiae* Gray 1826, *Canis familiaris novaehollandiae* Voigt 1831, *Canis dingoides* Matschie 1915 and *Canis macdonnellensis* Matschie 1915. In 1956 the earliest scientific name of the dingo, *Canis antarcticus* Kerr 1792, was suppressed in Opinion 451 of the International Commission on Zoological Nomenclature in favour of the more popular *Canis dingo* Meyer 1793. In 1982 the specific designation *Canis lupus* was recommended over *Canis familiaris* on the basis of universal usage, even though the latter name had page priority. Thus, the current scientific name of the dingo is *Canis lupus dingo*.

This name not only implies that the dingo is a single subspecies of the grey wolf, but dingo populations are uniform throughout their huge distribution in Asia and Australia. Recent evidence (below), however, indicates the existence of regionally distinct populations which may represent subspecies.

Morphological comparisons of dingoes in Australia and Thailand

Samples from different regions in Thailand (Tharae, Chiengmai, Ba pa Miang) plus samples from the same region (Tharae) in different years were compared by making a canonical variate analysis of skull variables. There was no significant difference between any group ($P<0.001$) to indicate that these Thai samples were identical between years and regions, which suggests that dingoes in Thailand belong to a single population. Similar comparisons of skull variables between samples of dingoes from central Australia and Thailand, however, were significantly different ($F=35.0$, $P<0.001$; see Figure 1.3), as were comparisons of Thai dingoes with 'dingo-like' canids from Israel, Japan and Korea. Also, Thai dingoes of both sexes are significantly smaller than Australian dingoes in seven standard body measurements (t-test, $P<0.001$).

The conclusion from these comparisons is that the Thai population is sufficiently distinct in skull and body morphology to warrant subspecific status, and henceforth should be known as the Thai dingo. An appropriate nomenclature might be *Canis lupus siamensis.*

Morphological comparisons of dingoes in Australia

The leading question is: 'Are regionally distinct dingo populations a reality that represent subspecies or is there only one population that varies in response to a set of environmental variables (clinal variation)?'

Research has clearly demonstrated that dingoes can be defined on the basis of skull morphology (Chapters 1 and 10). 'Pure' dingoes (defined this way) are subdivided into 'statistically distinct' regional populations in northern, central and south-eastern Australia (Figure 3.3). These data support the concept of alpine, desert and tropical races of Australian dingoes, and so do differences in size, coat colour and some reproductive data.

On the other hand, much of the data overlaps the groups; for example, skull scores for many south-east dingoes fit well into the range for northern populations and vice versa. Further, if similar statistical testing is conducted on each

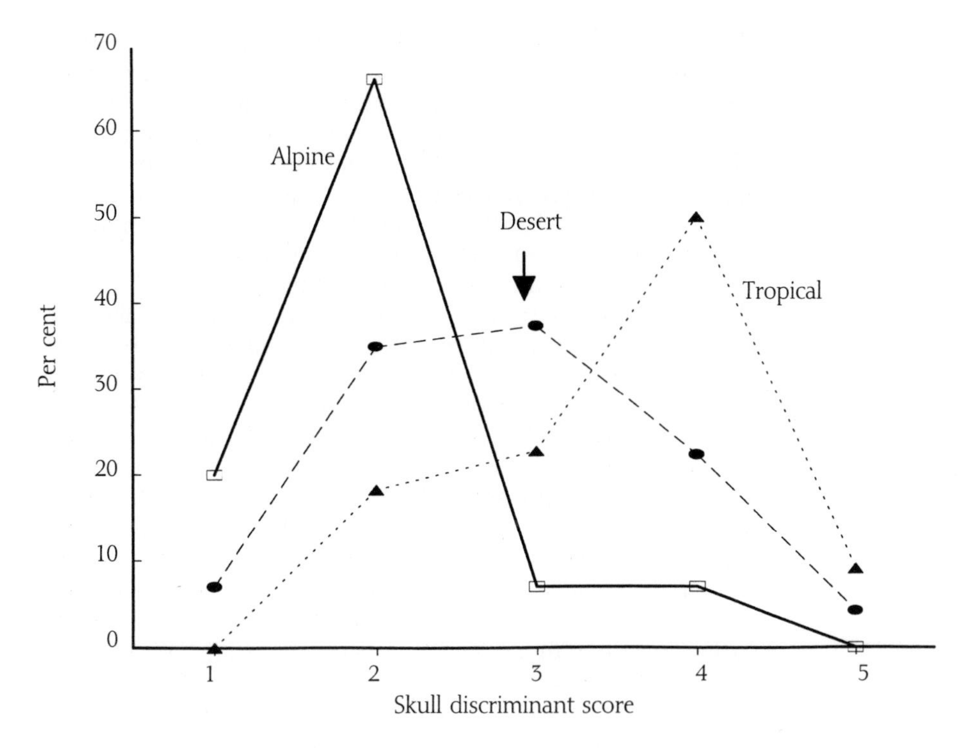

Figure 3.3

Alpine, desert and tropical dingoes: fact or fiction? The graphs represent the percentage of dingoes in each category (1–5) of skull discriminant scores. Alpine, desert and tropical dingo samples are from south-eastern (lat. 37°S; mean skull score ± sd: 1.55 ± 0.64); central (lat. 23°S; 2.45 ± 0.85); and northern (lat. 13°S; 2.92 ± 0.81) Australia respectively. Global tests of the mean score for these regions, using various statistical procedures, all indicated significant differences between dingo populations in south-eastern, central and northern Australia (south vs central $P<0.001$, south vs north $P<0.001$, central vs north $P<0.05$). The scores were derived from equation 7 in Newsome & Corbett, 1985 (see also Appendix E).

regional group — central populations, for example — then dingoes in the Harts Ranges (23°S) statistically differ from those on the edge of the Simpson Desert (26°S). In fact, there is a significant correlation between skull discriminant score and latitude for six dingo populations sampled at 13°S, 14°S, 19°S, 23°S, 26°S and 37°S ($r = -0.81$, $P<0.05$). These data suggest that there is clinal variation along a climatic gradient, perhaps one involving temperature and rainfall.

So where does one draw the line? How far does one split hairs? An analogy may be useful here. Suppose a Martian spaceship observed humans from an orbit high over the earth; the crew might conclude that humans could be divided into subspecies on the basis of colour (black, white, yellow etc.). However, on hovering closer over a particular region, say Australia, and seeing humans of various colours living more or less together, the Martians might doubt their previous conclusion. Then, should the spaceship land, an astute Martian would pick up differences between humans of the same colour, even between those living in the same house. The Martians might then conclude that there are about 17 million subspecies of humans in Australia alone!

For dingoes, an analysis of the existing data bank is equivalent to the Martian view of humans from afar: it looks as if there are relatively homogeneous populations in three regions of Australia, and thus three valid subspecies appropriately named alpine, desert and tropical dingoes. However, an analysis of further samples (skulls or DNA) from other geographic regions is required to confirm that conclusion. If it is confirmed, and given that dingoes were first described from Port Jackson, New South Wales (the type locality), the correct nomenclature for the Australian alpine dingo would be *Canis lupus dingo* Meyer 1793. The earliest name for central Australian populations was established in 1915, so that the correct scientific name for the Australian desert dingo would be *Canis lupus macdonnellensis* Matschie 1915.

There appears to be no existing scientific name for northern populations. Dingoes were frequently recorded by the first European settlers in tropical Australia (1838–49) at Port Essington on the Cobourg Peninsula of Arnhem Land. Since the first scientific specimen also appears to have been collected from Cobourg Peninsula (in 1965), an appropriate name for the Australian tropical dingo might be *Canis lupus cobourgensis*. This would need to be formally submitted to and accepted by the International Commission on Zoological Nomenclature.

CHAPTER 4

LIVING AREAS AND MOVEMENTS

This chapter describes the spatial organisation (living areas) of dingoes. How large an area do dingoes need to survive and raise offspring? What are the major determinants of living area size in different regions of Australia — prey characteristics (type, abundance, catchability), habitat (topography, availability of water) or dingo density?

TYPES OF LIVING AREAS: HOME RANGE AND TERRITORY

For carnivores, *home range* implies that an animal is resident in a particular area for a relatively long time period, usually years. It is defined as the area over which an animal normally travels in pursuit of its routine activities such as hunting, meeting conspecifics and raising offspring.

A *territory* is defined as the area in a home range that is occupied more or less exclusively by an animal (or the individual home ranges of a pack of animals), and is maintained by overt defence or advertisement. For dingoes, it appears that only packs defend territories throughout the year.

As described in following sections, the home ranges of neighbouring dingo individuals and pack territories sometimes overlap spatially to some degree, especially during the breeding season or in areas where resources are shared, such as water holes or hunting grounds. In shared areas there is usually minimal temporal overlap by neighbouring individuals (or packs) in communal areas.

Some individuals, described as loners, or vagrants, temporarily cohabit the home ranges and territories of other resident dingoes but they rarely associate with them except for evasive or aggressive interactions.

Table 4.1

Home range size (mean and range in km^2) in contrasting habitats in Australia. Size of home range was estimated by the minimum convex polygon method. The mean range uses all available data and was calculated (or estimated for the Fortescue data) to directly compare home ranges in all habitats, because for some regions it was not known whether or not data included information on juveniles and loners. For location, the numbers in parentheses refer to study sites given in Table 2.1.

Location (study site)	Habitat	Mean range	Males	Females	Loners	No. of dingoes
Fortescue River, North-west Australia (3)	Semi-arid, coastal plains & hills	77	61	52	160	19
Simpson Desert, Central Australia (6)	Arid, gibber & sandy desert	67 (32–126)	98 (70–126)	47 (32–68)	?	5
Kapalga, Kakadu NP, North Australia (1)	Tropical, coastal wetlands & forests	39 (15–88)	36 (15–52)	42 (22–88)	?	18
Harts Ranges, Central Australia (4)	Semi-arid, river catchment & hills	25 (7–110)	17 (7–54)	17 (7–44)	55 (27–110)	24
Kosciusko NP, South-east Australia (9)	Moist, cool forested mountains	21 (2–54)	24 (6–54)	19 (2–53)	?	13
Georges Creek NR, East Australia (8)	Moist, cool forested tablelands	18 (4–55)	20 (5–55)	12 (4–19)	?	8
Nadgee NR South-east Australia (11)	Moist, cool coastal forests	10 (9–14)	10 (9–10)	11 (9–14)	?	5

A COMPARISON OF LIVING AREAS IN CONTRASTING HABITATS IN AUSTRALIA

In general, the largest recorded home ranges occur in the arid regions and the smallest are in the moist forested mountains of eastern and south-eastern Australia (Table 4.1). Loners have larger ranges than residents, but only 2 of 8 studies distinguished these categories. Except for Kapalga, males tend to occupy larger ranges than females, doubly so in the Simpson Desert.

Fortescue River region, north-west Australia

The best data come from Thomson's 9-year study in the Fortescue River region which also provides the only data on pack territory size (Table 4.2) and space (Figure 4.1).

Individual home ranges and pack territories were both roughly ellipsoidal in shape. The five resident packs all displayed strong site attachment and their territorial boundaries were quite stable from year to year. On 19 occasions when researchers were radio-tracking pack members in the overlap zone between two adjacent pack territories (B and E), the

Table 4.2

Dingo territories and resources availability in the Fortescue River region, north-west Australia (adapted from Thomson, 1992).

Dingo density is the mean number of dingoes per 100 km². Large kangaroos (euro, red kangaroo) are the major prey of dingoes in this region and an index of their abundance is expressed as the percentage of observations of kangaroos per observations of dingoes. The mean territory size was 80 km².

Pack	Territory area (km²)	Mean pack size	Dingo density	Index of kangaroo abundance	Habitat, % of territory (% used by dingoes)			
					Riverine	Stony	Floodplain	Hills
A	113	12	11	15.9	10 (49)	1 (2)	21 (6)	69 (44)
B	94	12	13	8.5	14 (43)	9 (10)	38 (25)	39 (23)
C	86	3	4	3.9	2 (3)	0 (0)	63 (94)	35 (3)
D	63	6	10	12.3	12 (35)	5 (8)	46 (20)	37 (37)
E	45	10	23	8.4	14 (31)	6 (4)	39 (18)	42 (47)

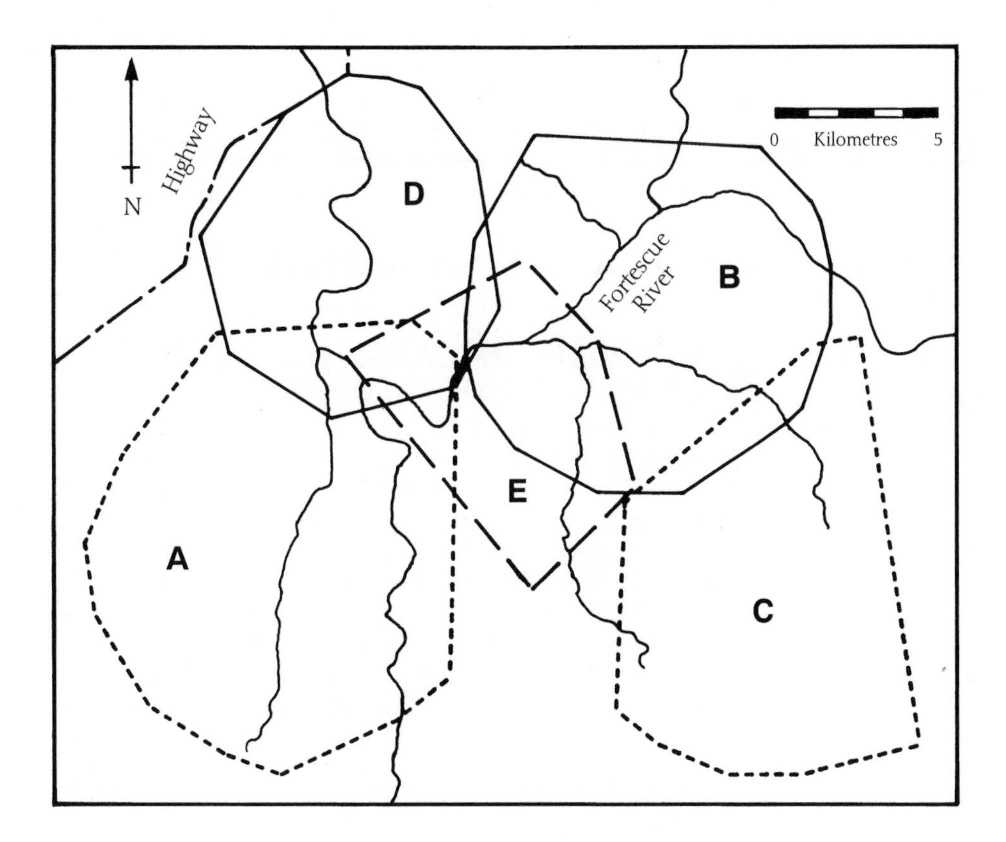

Figure 4.1

The spatial relationship between five stable dingo pack territories (A–E) in the Fortescue River region in north-west Australia (adapted from Thomson, 1992). The convex polygons represent the territorial boundaries determined by 4,194 independent radio-tracking locations over 4 years. The territories' sizes are given in Table 4.2.

mean minimum distance between members of opposite packs was 2.6 km (range 0.3–4.9 km). Even during the 9 recorded occasions when members of one pack were in the territory of the other pack, the average minimum distance between intruders and residents was 2.9 km (range 4.7–11.6 km).

In the Fortescue region, no clear relationship was found between territory size and any single important factor such as pack size, dingo density, prey availability, habitat type and the use made of each habitat (Table 4.2). The largest pack (A) occupied an area similar in size to the area occupied by the smallest pack (C). This pack (C) also carried fewer dingoes (lowest dingo density) and fewer kangaroos (lowest prey index), and occupied the lowest proportion of prime riverine habitat. The pack with greatest dingo density (E) had medium prey availability and one of the highest availabilities of riverine habitat. Dingo use of the various habitats was disproportionate to availability. Riverine areas (rivers and creeks containing water and associated thick vegetated cover) were intensively used by packs whereas stony areas were used infrequently. These data suggest that the size of most living areas is determined by a tradeoff between dingo density and the availability of food and water resources.

Kapalga, Kakadu National Park

Individuals and packs alike occupied relatively large, essentially non-overlapping living areas which usually encompassed both forest and floodplain habitats. The temporal use of living areas alternated between the wet and dry seasons, simply because of accessibility. For example, one adult radio-collared male occupied an area for 33 months but lived almost exclusively on the open floodplains in the dry season (camping in mangroves by a river), then shifting to the forest end of his range during the 3-month wet season.

During the dry season, many dingoes live on the floodplains and hunt rats, geese and grasshoppers. The separation of individuals of rival packs appears to be based on frequent observation of one another and mutual avoidance (see Chapter 6). In contrast, adults that are resident in forest habitats use their total area more evenly and probably more exclusively. For example, one adult female regularly moved around the full extent of her forest range for 19 months before radio contact was lost.

Georges Creek Nature Reserve, east Australia

The elongated (ellipsoid) shape of home ranges in this deeply dissected escarpment habitat were greatly determined, it seems, by the orientation of the valleys and ravines. Dingoes used some parts of their home range frequently (core areas) and other parts rarely. In frequently used areas dingoes made searching movements in one area for a time, then moved to another area and established a new pattern of searching movement. Areas of searching movements were joined by exploratory movements (see below). Some of the areas that dingoes searched (probably hunting areas) were also used by up to four other dingoes, but at different times. Although there was much overlap of individual living areas and communal use of many resources, including hunting grounds and den sites, dingoes nevertheless tended to avoid each other by temporal separation when they used shared resources.

Synthesis of living areas: reasons for regional differences

The smaller size of the home ranges of dingoes in the eastern highlands (Table 4.1) is probably due to high productivity in the relatively moister habitats there, the milder climate and the relatively larger diversity of prey. Even so, it is the presence and distribution of resources that mostly influences the disposition of dingo packs in their living areas. In hot arid areas the distribution of water is clearly the single most important factor that determines where dingoes and their prey can be found, but food resources will ultimately dictate how long a pack of dingoes can forage in any single locality.

The most important factor determining living area size is, therefore, a reliable food source (rather than abundance per se), and perhaps its regularity (predicability). In arid Australia there are usually boom and bust periods (flush and drought) where small prey form plagues for several months then disappear for years. In the tropics, similar events occur, but usually each year. Dingoes therefore need large living areas to exploit such relatively irregular (unpredictable) food resources. However, in the temperate forests where food is more diverse and reliable, living areas can be smaller. Although occasional forest wildfires may change the frequency of different prey species, food is always plentiful, and so is water.

SHORT-TERM MOVEMENTS IN LIVING AREAS

Fortescue River region, north-west Australia

The mean minimum distance travelled each day by radio-collared dingoes was 3.3 km (range 0–12.9 km) for pack members and 5.6 km (0.2–18.6 km) for lone dingoes. Actual distances travelled were probably much greater because most animals were not radio-tracked continuously; they may have started and finished at the same point but travelled a considerable distance in the meantime. For example, one dingo returned to its starting point after travelling at least 19.6 km in 7.5 h. Since the mean maximum width of ranges of packs in the Fortescue was 10.5 km, these data indicate that most movements were local and that pack dingoes did not regularly patrol the boundaries of their territories.

Pilbara district, Western Australia

The movement patterns of juvenile dingoes were studied in this habitat, which consisted mainly of spinifex steppe with scattered low hills and granite outcrops, and was dissected by dry sandy riverbeds with intermittent water. A total of 108 pups (4–8 weeks old) were tagged and released at natal den sites. Eleven months later the distances between tag and recapture sites varied between 0.8–34.4 km. The mean distance for male and female juveniles was 21.7 and 11.0 km respectively. These data suggest that males disperse more than females do, which is consistent with similar data from other regions.

Harts Ranges, central Australia

All home ranges were centred on water (artesian bores or dams) and were approximately circular. Radio-tracking clearly indicated that females had fixed home ranges centred on their birth areas and did not disperse far; the longest recorded distance was 9 km from one valley to the next. For example, one female, first captured as a pup, was recaptured 2.5 km from her birth area 36 months later, and again 45 months later as a mother with pups. The following year she whelped in the same valley and the next year she was still operating only 3–4 km from her natal den area, 73 months after initial capture. Similar data were obtained for another 6 females (maximum 5 km movement from initial capture point up to 36 months later).

The movements of most males were similar. For example, one resident adult lived in a fixed home range for 79 months with a maximum movement of 9 km; another adult male moved from one valley to the next soon after, possibly as a result of being captured, and remained there for at least 54 months. The longest recorded movement was 32 km by two males, probably because their water hole dried up and they were forced to disperse. However, since radio contact was lost with males more often than lost with females, it is possible that males generally made longer movement patterns than females did.

Simpson Desert

Most home ranges were elliptical and most movements within them were aligned along creeks. One female moved 15 km in 48 months, another female operated for 11 months in one area; when the water dried up she moved 23 km to the next water. No data are available on the short-term movements of males in this region.

Georges Creek Nature Reserve, east Australia

Overall, dingoes in this region had small home ranges and limited movements. Radio-tracking indicated two types of movement patterns. *Searching movement* was characterised by intense activity in a small area and large, angular, frequent changes in direction; it also appeared to be associated with hunting. *Exploratory movement* was characterised by apparently more purposeful movement from one place to another, and the traversal of a substantial area. Exploratory movements either joined two separate searching movements in different localities or served as a loop between two searching movements in the same locality. In both cases, many exploratory movements followed the boundaries of home ranges, and seemed related to maintaining communication between animals. The dingoes may have been visiting and maintaining scent posts along ridgetops, creeks and firetrails.

Dingoes of both sexes were active throughout most of the 24-h day with peaks of activity at dawn and dusk, and lowest activities were at noon. Exploratory movements tended to be more frequent between sunrise and 1100 h and between 2200–2400 h than at other times. Searching movements occurred throughout the 24-h day with a peak around sunset. Activity periods were typically short (mean duration 52 ± 11 min) and separated by shorter rest periods (70% of rest periods were ≤30 min, maximum 3 h). On average, dingoes were active for about two-thirds (15.25 h, 65%) of each day. The mean daily distance travelled was about 16 km. The distances travelled per hour in diurnal and nocturnal periods were equal. The mean continuous distance travelled was 1.07 km and the longest continuous movement recorded was 9.8 km over 5 h. The mean rate of movement was 1.3 km/h and 1.1 km/h for adults and juveniles respectively.

These data suggest that dingoes caught prey (mainly wallabies; see Chapter 7) by stealth or surprise rather than by fast, long pursuit, which would have been difficult in the broken terrain and dense vegetation.

Kosciusko National Park

Home range shapes were elongated (ellipsoid) and the longest axis was 12.2 km. The mean maximum distance moved between successive radio-tracking days was 5.8 km and 3.6 km for males and females respectively; and the overall average was 2.0 km. The longest recorded movement was 20 km. These data suggest that dingoes in this mountainous region had limited movements, consistent with dingoes living in similar escarpment habitat, as described above.

FORAYS AND DISPERSAL FROM LIVING AREAS

Fortescue River region

Resident dingoes seldom travelled far or for long periods beyond the boundaries of home ranges or territories. Such forays were usually conducted by individual males (68% of 22 forays), sometimes small mixed groups (mean size 2, range 1–5), and involved distances of 2–10 km beyond the boundary (most within 4 km of boundaries). The mean duration of forays was about 24 h (range 1–72 h) which was similar for solitary dingoes and groups. Juveniles rarely made forays. Dingoes were not observed to engage in any particular activity during forays, but since most forays were made by males, they might have involved territorial behaviour such as detecting intruders or assessing the occupancy of neighbouring areas. Some forays may have been 'apparent' departures from territories, reflecting movements of dingoes close to their artificially computed boundaries. Few forays seemed to be related to the breeding season or any other part of the year, but some may have been related to subsequent dispersal, as indicated below.

Dispersal movements — permanently moving out of the original natal territory — were recorded on 31 occasions over 9 years, and involved 25 individuals (pack members and loners), five small groups and one entire pack. Individual males dispersed not only more often than females did but over longer distances. This pattern is typical of all canids except African hunting dogs where females disperse and males remain with their natal pack.

Adult dingoes tended to disperse more frequently than juveniles did. The mean dispersal distance for individual dingoes (males and females combined) was 20 km (range 1–184 km), and 9 km (3–16 km) for the groups. Dispersal was usually a gradual process, with most dingoes dissociating from other members of the pack (either socially or spatially) for varying periods before leaving the territory; others left the pack territory without any detectable changes in movements or behavioural associations before dispersal.

Dispersal was recorded throughout the year. About half of the dingoes that dispersed managed to settle into a new range, choosing either a vacant area or associating with the residents in an occupied range. In the former case, most dispersing dingoes ended up at the same site, a 'dispersal sink' created by human control activities such as trapping, poisoning and shooting. This site attracted them because of the absence of resident dingoes, which meant abundant resources, but most (16/24) dingoes that dispersed there were eventually killed by human control activities (see Chapter 8).

Dispersal occurred at all levels of population density but was highest when population density was high and food supply was low, and was determined by the number of vacant areas available. As for wolves, food resources in conjunction with territory vacancy ultimately set the rules for the dispersal strategy for dingo pack members.

Simpson Desert

Radio-tracking suggested that dingoes, particularly males, regularly travelled long distances, and that most movements were made along dry creeks. For example, 8 males moved between 36 and 250 km in 3 weeks to 10 months. As for the Fortescue River region, the relatively high rates of dispersal in this region were thought to be related to human control activities that created vacant areas (dispersal sinks). In contrast, the longest dispersal movements by males occurred after exceptionally good rainfall had stimulated the eruption of plagues of small mammals. In these circumstances, movements were made across the gibber plateaus and sand dunes between the creeks, rather than along creeks.

The data in this chapter clearly indicate that the separation of most dingo individuals and packs is essentially spatial but temporal separation occurs when resources are shared. This temporal separation is quite important because of the close proximity of rival individuals and packs and the potential for fighting and its consequences. How do dingoes avoid each other? How do rivals know when to intrude into their neighbour's domain and yet avoid contact? Surely residents must know what is going on? Lone dingo living areas greatly overlapped those of packs but in this case, also, few encounters between them were recorded. The next two chapters deal with these and other questions about how dingoes operate and maintain their living areas.

CHAPTER 5

BEHAVIOUR AND COMMUNICATION

This chapter describes aggression, howling and scent-marking — the behaviours that are most important in maintaining the cohesion of packs and their living areas so that resources can be used efficiently.

DEFINITIONS AND DESCRIPTIONS OF AGGRESSION, DOMINANCE, SUBMISSION AND OTHER SOCIAL BEHAVIOUR

The following behaviours were observed in both captive and wild dingoes.

Aggression and responses to it

The typical posture of a threatening dingo is an upright body stance, stiff legs, a raised tail often curled over the back, and the teeth bared while snarling. The white tail tip most dingoes have probably flags the tail position of a dominant dingo.

Low-intensity threats comprise staring intently either in silence or snarling with teeth bared. In response the threatened dingo turns or lowers its head in an attempt to avoid eye contact. More intensive threats involve close or actual body contact accompanied by much snarling and pronounced teeth-baring. Lunging and snapping is usual but biting is uncommon and inhibited. Contact involves mounting on the shoulder or hips or from behind, backing into, hip-slamming, or standing over or on the threatened dingo. The threatened dingo responds with *passive submission* and either crouches, flattens on the ground, or rolls over on its back with typically slow, deliberate movements. It holds its ears laterally or flattened and keeps its tail down or tightly between its hindlegs and against its belly. The lips usually stretch backwards, without teeth-baring, in a so-called 'submissive grin' and the tongue often protrudes; otherwise there is pronounced teeth-baring. Mounting, standing over and genital licking by the threatening dingo is allowed (see Plate 14).

Outright attack leading to serious fighting and the wounding of combatants either occurs spontaneously or ensues from less intensive threat behaviour or from a returned threat. Otherwise the threatened dingo squeals and quickly rears back or flees.

The response to threat behaviour by oestrus and lactating females is defined as *tolerance behaviour*. Instead of passively submitting or returning the threat, the recipient male stands in a semi-threat posture without snarling, then averts his head and moves away.

Active submission

Movements are typically fast and involve one or several participants. The participant goes in front of a dominant elder, usually the alpha male or female (see below), in a slightly crouched posture, ears flattened and tail down and wagging. It licks the muzzle of the dominant dingo and often mimics its movements. In response the dominant dingo stands aloof in a typical threat posture or initiates *following behaviour* or other friendly behaviour. Aggressive interactions and vocalisations are rarely part of active submission.

Friendly behaviour

Examples include play, play fighting (with inhibited biting), greeting, grooming other dingoes and resting close together. *Family behaviour* includes parental behaviour and alloparental (helper) interactions.

Social rank, dominance and the pack

Packs are essentially extended families, similar to wolves and other canids; they comprise a mated pair, their offspring of the year plus some offspring from previous seasons. The nature and frequency of social interactions between the adult group members of a pack vary greatly. Even so, there is always a rank order or hierarchy whereby some dingoes are more likely to win disputes than others. In a stable hierarchy a dingo that consistently initiates aggressive behaviour and wins disputes is defined as dominant to its adversary, and the latter is defined as subordinate and performs passive submission.

The absolute dominant dingo is termed the alpha (rank 1).

Others immediately lower in the hierarchy are arbitrarily termed high-ranking (or beta, ranks 2–3) and those at the bottom of the hierarchy are arbitrarily termed low-ranking (or gamma, ranks 4–≥6). This group of interacting adults, with their weaned pups, is defined as a pack. Some of the lowest ranking adults, usually young animals, do not run with the pack and are termed outcasts or loners.

THE IMPORTANCE OF DOMINANT BEHAVIOUR

In a study of captive dingoes in Alice Springs, very young pups were raised in isolation from their parents and siblings and, at various ages, tested against pups (of similar age) that had been raised in normal dingo society. The results were clearcut in that all the 'isolated' pups performed only aggressive behaviours whereas the 'normal' pups performed aggressive and/or submissive behaviours depending on their experiences and social status. Even when an 'isolated' pup was being mauled by a 'normal' aggressive high-ranking pup, the 'isolated' pup persisted with aggressive behaviours. These laboratory experiments indicate that dominant behaviour is innate (inherited) whereas the expression of submissive behaviour is learned from other pack members.

This was supported by field observations when a 3-month 'isolated' pup (named Hercules) was released and allowed to interact with a litter of five wild dingoes of similar size and age. After initial aggressive interactions between Hercules and the wild pups, the latter seemed confused by the behaviour of an animal that did not submit when he was 'supposed to'! After two days, Hercules displayed no submissive behaviour (essentially because he did not know how to), and became the leader of the group; the wild pups followed his movements and usually submitted passively whenever they made direct contact.

The parents of the wild litter made no contact with Hercules, probably because of the presence of the human observers. However, as suggested by other data from this region of central Australia, when different wild litters became mixed, they probably would have accepted him (see Chapter 6).

VOCALISATIONS AND THEIR FUNCTION

Several kinds of vocalising have been observed in dingoes: howls (long-range communication), moans (short-range communication), bark-howls and snuffs. Other canids similarly have a wide repertoire; for example, 11 categories of vocalisations have been recorded in coyotes.

Howling

Howling is the most frequent form of vocalisation and there are three basic types with at least 10 variations (Figure 5.1). The chorus represents a group of individuals howling together where each uses a different variation of the basic howls. Pup howls are usually discernible to a human ear by their relatively higher pitch, and by either the shorter duration of plateau howls or an abrupt cutoff after the maximum pitch is reached. The following analysis of howls is supported by the analyses of spectrograms summarised in Box 5.1.

Howling has one basic function — to determine the *location* of other dingoes — but it serves two apparently conflicting purposes: (1) to *attract* and unify other members of the pack, or to help loners find breeding partners; and (2) to *repel* rival packs or individuals, thereby defending territorial resources and/or reducing social friction.

To distinguish between howls that are meant to attract and those that are meant to repel, dingoes must be able to identify the howler(s), and different howl types must have specific, reliable messages. It has not been scientifically determined whether each dingo has its own, individual 'voice' (as has been suggested for wolves), or whether dingoes can count or otherwise deduce the number of howlers, but anecdotal evidence supports these two possibilities. The main characteristics of howls (type, frequency) appear to vary according to factors such as the proximity and 'value' of an essential

resource such as water and food, the breeding season, experience, age, sex, and social status. However, to identify the howler(s), dingoes must be able to take into account other factors such as their sighting of the howlers, the immediate prior locations of rivals, and information from pheromones (scent posts).

Box 5.1
Spectrograms of howls of wild dingoes living in the Simpson Desert

A howl is a tonal sound with three distinct properties — frequency (pitch), amplitude (loudness) and timbre (quality), and the spectrographic analysis of a sound wave is based on these three properties.

The shape of the fundamental band of dingo plateau howls is flat and it has very little frequency variation. The howl begins at a low pitch, rises slightly and remains constant almost to the end before dropping slightly. The harmonics for plateau howls range in number from 1 to 6 (mean 2). The formant (the envelope of the frequency pattern arising in the vocal cords which leads to the aural recognition of sounds) is usually the first harmonic and the overall frequency of harmonics ranges from 138 hz to 4,200 hz. The mean starting pitch (frequency) of plateau howls is about 540 hz and the mean duration is 2.9 s (range 0.3–5.2).

Chorus howls are characterised by a series of steady, flat and low-pitched howls that form the basis of the chorus. Overlaying these are very high-pitched modulated howls, which usually superimpose the harmonics of the low flat howls to form complex waves with greatly variable contours. For example, some are very modulated at the beginning, then flatten out and maintain a steady pitch throughout the chorus. Others are short, repetitive and very high-pitched bursts that sound like 'yip-yaps' to the human ear. The number of harmonics ranges from 1 to 6, averaging 2 for the low-pitched howls and 5 for the high-pitched howls. The formant is usually the first harmonic, sometimes also the second harmonic. The mean starting pitch for individual howls is 895 hz and the mean duration of individual howls in a chorus is 3.0 s (range 0.8–6.3). All dingoes howl at very distinct pitches during a chorus, and no two dingoes have been recorded howling at the same pitch during the chorus. As the number of animals joining in the chorus increases, so too does the variation in pitch.

Bark-howls commence with a short abrupt bark followed by a sudden rise in frequency and amplitude, lasting 0.2–0.5 s, then continuing flat and steady at a lower frequency during the vocalisation.

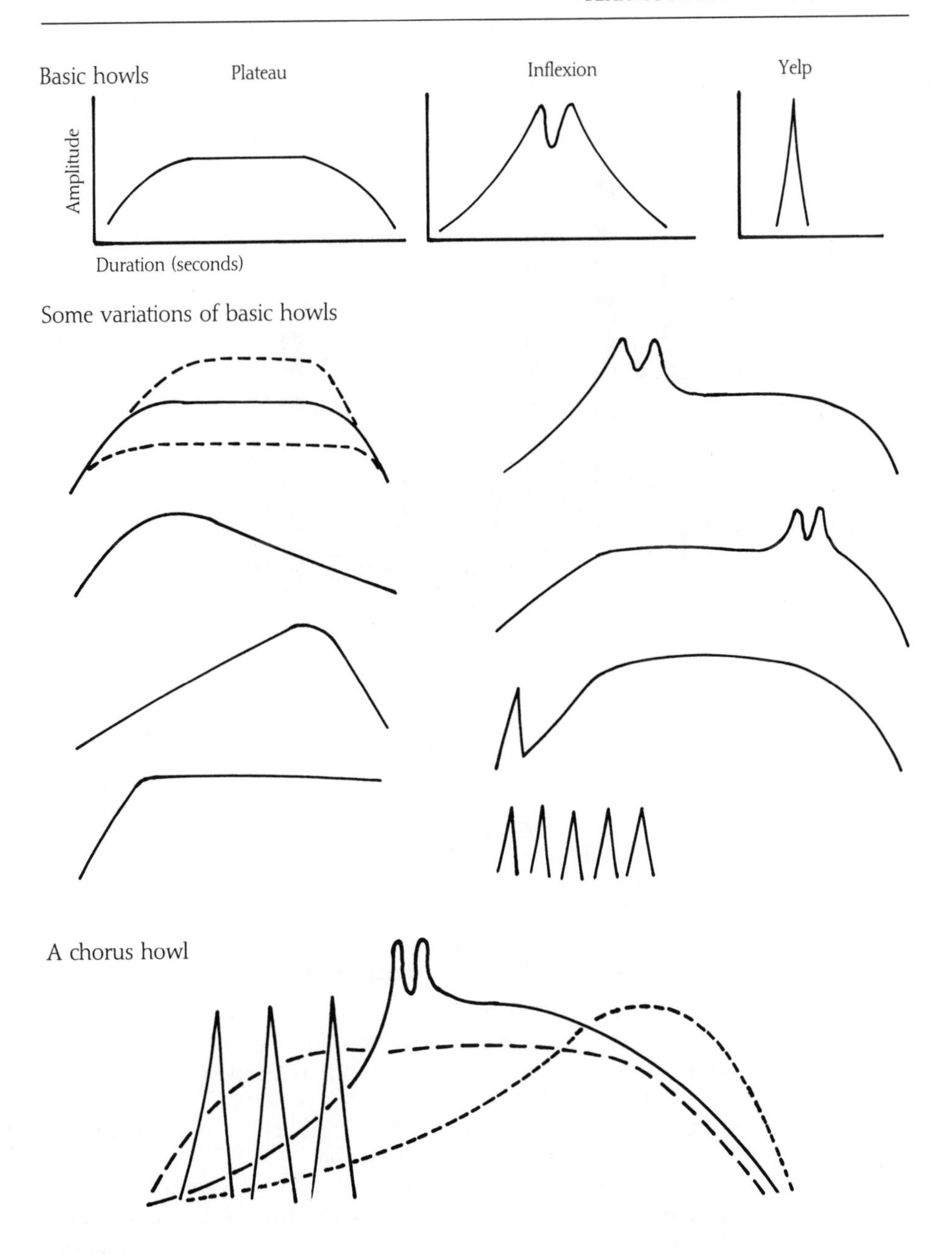

Figure 5.1
The three basic types of howls: plateau (holding one note), inflected (yodel), and abrupt (yelp). They may have an even rise and fall, an abrupt rise and protracted fall, the reverse, or a rise held as a brief high note. They are represented by relative amplitude vs relative duration (cf. spectrograms, Box 5.1), which is how they sound to a human ear. The figure shows a chorus howl by 4 dingoes, each using a different howl type; the number of howlers is clearly discernible to a human observer, and probably also to a dingo.

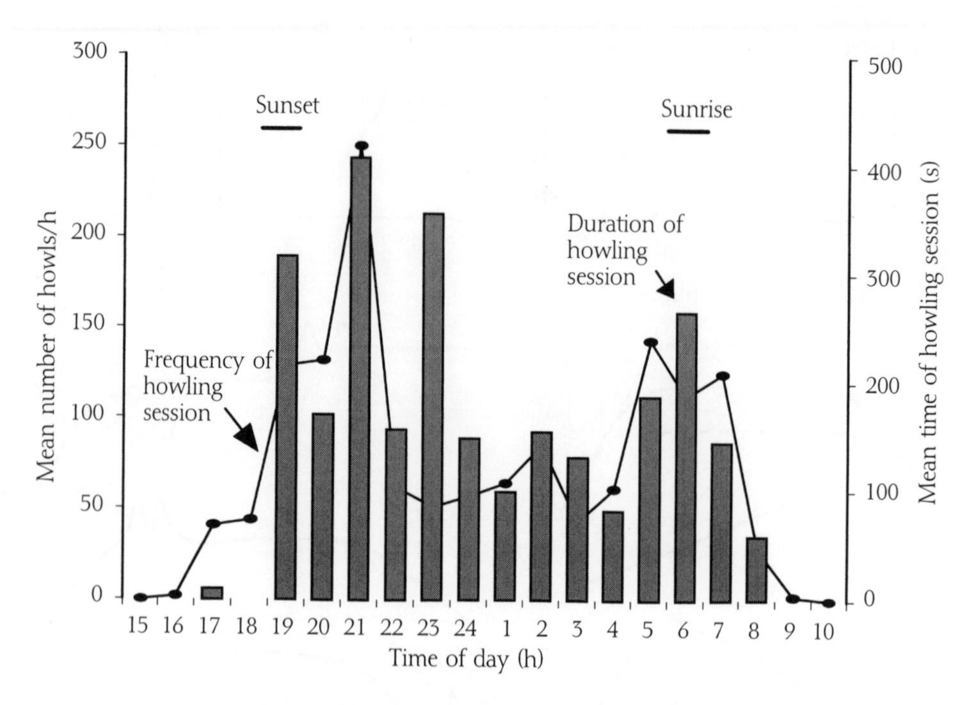

Figure 5.2

Daily howling by Simpson Desert dingoes, showing time of day when dingoes howled most frequently and duration of howling sessions. (5,920 howls were recorded over 12 months, mostly during crepuscular (dusk and dawn) periods and at night. Howling sessions were arbitrarily defined from when the first dingo began howling to when the last one finished responding; this involved 1–100+ individual howls. The mean duration of a howling session was 192.3 s.

Dingo packs in the Simpson Desert

Because dingoes living in the Simpson Desert are relatively undisturbed by human activities and run in stable packs throughout the year, their behaviour is probably representative of the ancestral or pre-European pattern. Observations of dingo howling behaviour, as described below, therefore reveal much about the function of howling.

Observations conducted over 12 months were based at an isolated watering point named 'Alka Seltza' (19 km and 10 km to the next waters) which was shared by about 23 dingoes. There were two stable packs (named East and West packs with 6 and 8 members respectively), 1–3 lone dingoes, and a group of 6 dingoes (named Desert Group) which arrived from the midst of the desert about halfway through the study and remained in the area for about 2 months, thereby increasing competition for both water and food.

Members of packs, groups and loners were identified by their coat colour, distinctive white body markings and scars.

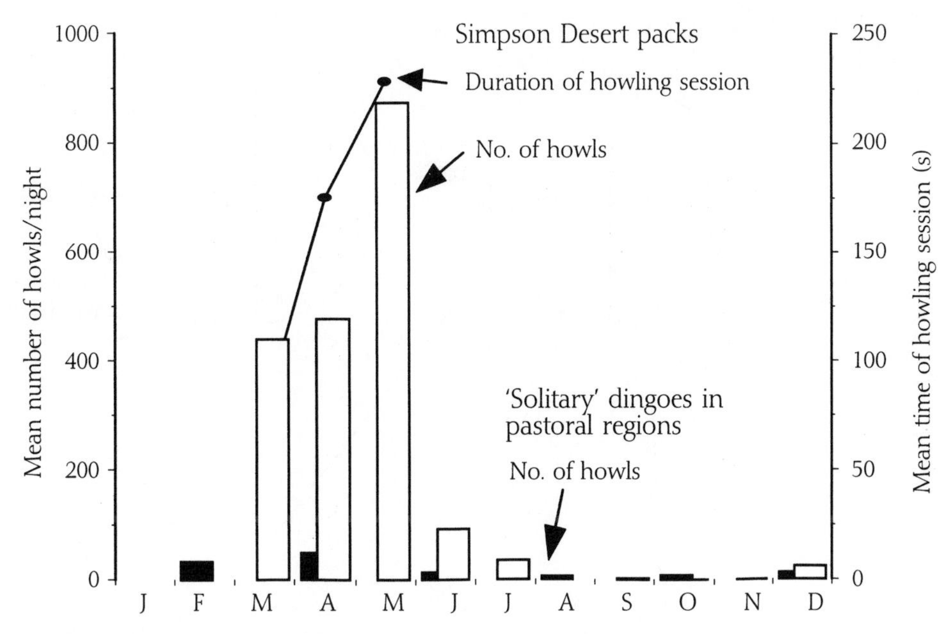

Figure 5.3

The seasonal frequency of howling by Simpson Desert dingoes (5,920 howls) compared to howling by dingoes in pastoral areas in central Australia (1,000 howls). One night was defined as 1700–0700 h. In the Simpson Desert howls per night ranged from 1 to 1,515. For both regions, most howling was linked to the breeding season. The line graph shows seasonal variation in the duration of howling sessions by the Simpson packs; there are no equivalent data for the others.

Interpretation at later periods was assisted by photographs and movie film. On many occasions, however, 'group' membership was assumed according to the area traversed by groups or individuals.

The main food source was rodents and rabbits whose populations fluctuated from low to high during the 12-month study. Sometimes this was artificially supplemented with carcasses (brumby, donkey, camel) placed at the water to heighten interactions between the packs. It is unlikely that this extra food distinctly altered howling or associated behaviour because howling was just as frequent with or without it. The dynamics of encounters and other interactions between dingoes at Alka Seltza are given in Chapter 6; only howling data is provided here.

Daily and seasonal howling sessions and howl types used by dingo packs in the Simpson Desert

All howling was done by the East and West packs (see Plate 16); neither the Desert Group nor the loners were ever heard

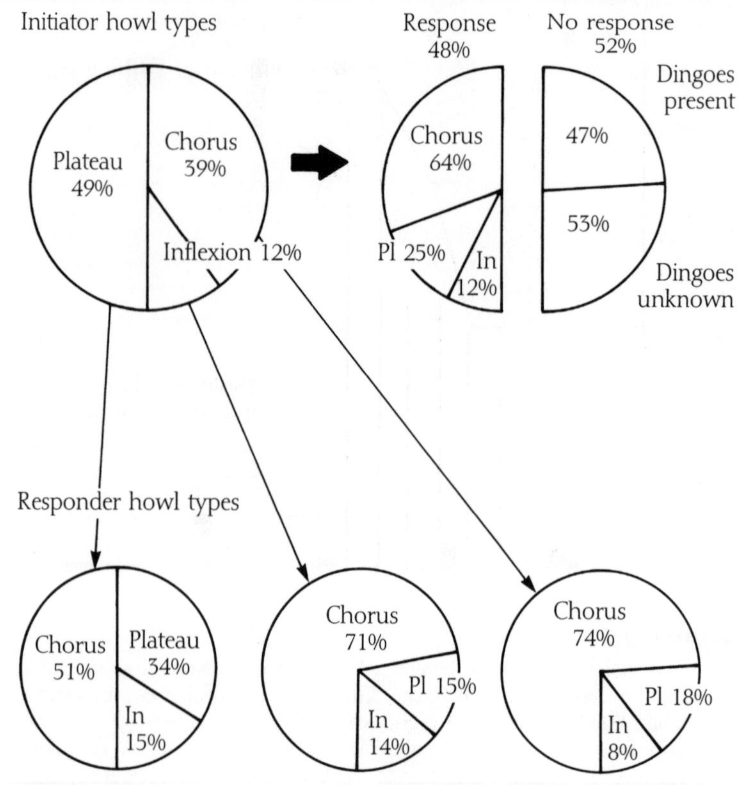

Figure 5.4

The types and frequency (%) of howls that initiated howling sessions (253 howling sessions), and the types (Fig. 5.1) and frequency of howls used by responding dingoes. Data for the East and West packs are combined to present an overall picture of dingoes living in the Simpson Desert. Pl = Plateau, In = Inflexion.

to howl, at least not near the water. Howling occurred throughout the night, but both the frequency and duration of howling sessions were greatest just after sunset and just before sunrise (Figure 5.2). Dingoes tended to howl relatively more often and longer in the dusk period.

The frequency and duration of howling sessions by packs were both greatest between March and June, which corresponded with the breeding season (Figure 5.3). By comparison, solitary and loosely associated groups of dingoes in pastoral areas elsewhere in Australia howled much less frequently, although there was also a highly seasonal trend linked to the breeding cycle.

Overall, dingoes that initiated howling used *plateau* howls in 49% of 253 howling sessions heard throughout the year (Figure 5.4). Dingoes responded to only 48% of these sessions even when it was known that they were often well within auditory range. Responding dingoes consistently used *chorus*

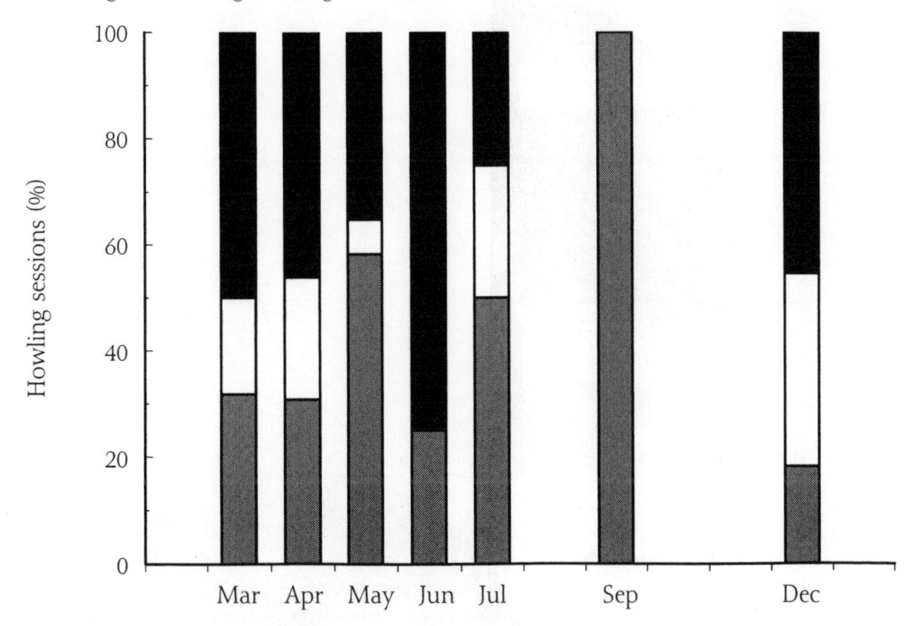

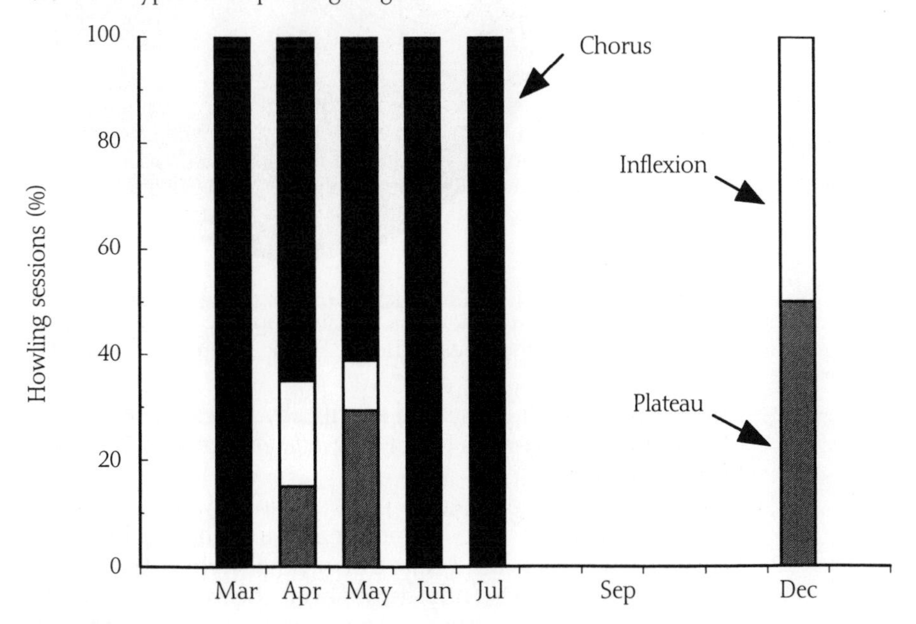

Figure 5.5

The seasonal variation in the type and frequency of howling by (a) dingoes initiating howling (253 howling sessions), and (b) dingoes responding to howls (122 howling sessions). Data were derived from dingoes living in the Simpson Desert.

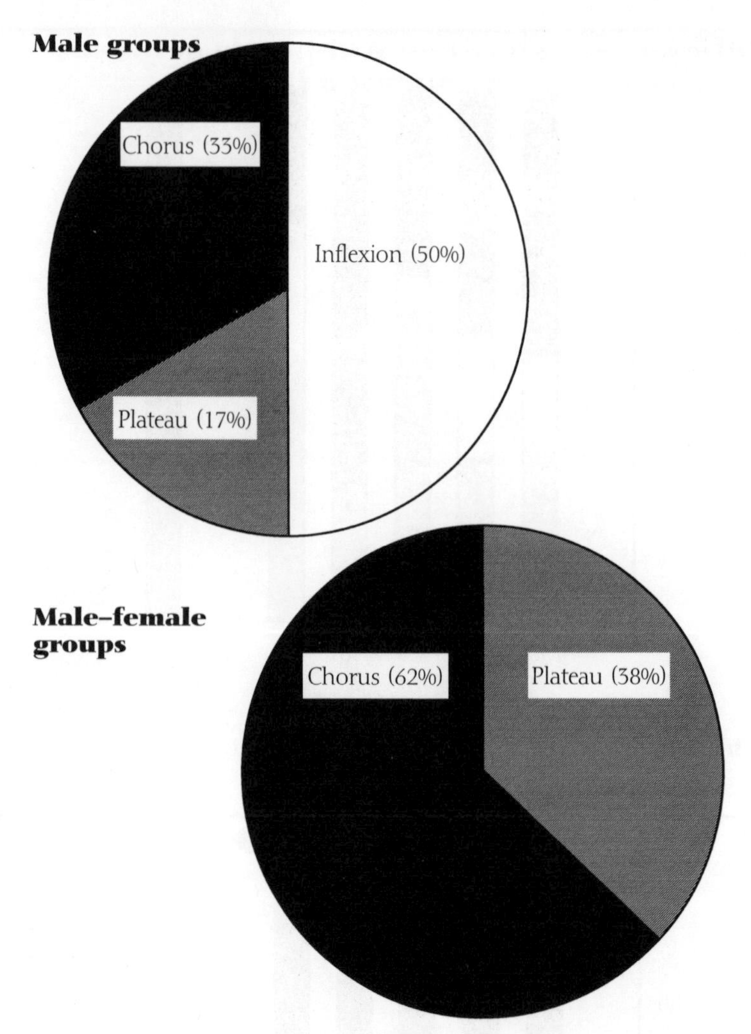

Figure 5.6

The percentage of chorus, inflexion and plateau howl types used by males and mixed groups in the Simpson Desert. Female groups did not howl.

howls, both overall (64%) and specifically in response to each of plateau (51%), chorus (74%) and *inflexion* howls (71%) used by the initiators.

There were also differences in the frequency of howl types used by initiators and responders throughout the year. Whereas initiators used all three howl types (especially plateau howls) consistently throughout the year, responding dingoes used chorus howls almost exclusively (Figure 5.5). The low frequency of howling between August and November was probably due to greatly diminished competition for food and water; massive rains had fallen in August, rodent populations resurged and dingoes focused on them.

Differences in howling within and between dingo packs in the Simpson Desert

There were major differences in the use and frequency of howl types within sub-units of packs. Male groups used all three howl types, especially inflexion howls (50%) whereas female groups never howled, and mixed male-female groups used only chorus (62%) and plateau howls (38%, Figure 5.6). Overall, males howled much more frequently than females did, but there was little difference in the types of howls they used and both used mostly plateau howls (91.7 and 85.7% respectively) rather than inflexion howls.

The alpha male and female of both packs frequently howled, and the alpha male of the West Pack initiated chorus howling on two occasions, but not enough data were collected to analyse social status and howl types because it was impossible to identify the dingoes' social status consistently.

The East and West packs were so named because they consistently used the hemispheres to the east and west of the watering point, which suggested that these were territories. There were clear differences between the type and frequency of the howls these packs used to initiate howling (Table 5.1), especially for inflexion and chorus howls. This may imply that both packs conveyed similar messages with plateau howls, but different messages for inflexion and chorus howls.

Table 5.1

The types of howls (%) used by packs to initiate howling sessions.

n = 30 and 59 howling sessions for the East Pack and West Pack respectively. A–D, E–G and H–J refer to variations in plateau, inflexion and chorus howls respectively, as shown in Figure 5.1. Neither the Desert Group nor the lone dingoes were ever heard to howl.

Plateau howl types	**A**	**B**	**C**	**D**
East Pack	26.7	3.3	3.3	0
West Pack	32.2	8.5	6.8	1.7
Inflexion howl types	**E**	**F**	**G**	
East Pack	6.7	6.7	10.0	
West Pack	3.4	1.7	3.4	
Chorus howl types	**H**	**I**	**J**	
East Pack	13.3	10.0	20.0	
West Pack	27.1	5.1	10.2	

Using howls to reinforce territories and gain access to shared resources

Pack members initiated most howling sessions when they were in their own territories and responded most when they were near the territorial boundary between the two packs (Table 5.2). For example, 69.7% of all howling sessions initiated by the East Pack came from east of the water and 44.4% of responses by the East Pack came from the west. The corresponding values for the West Pack were 77.8% and 31.9%. For most howling sessions, the identity of most dingoes that initiated or

Table 5.2

The frequency (%) of howling sessions according to the direction of howls recorded at the central observation site at water. Although howling came from 13/16 points of the compass, pack members initiated most howling from their own territories and responded most when near the territorial boundary. I = initiator, R = responder

	% Howling sessions					
	I		**R**		**I**	**R**
Direction of howls	East pack	West pack	East pack	West pack	Unidentified pack	
North	–	–	–	2.1	5.0	3.3
NNE	–	–	5.6	2.1	–	0.6
NE	9.1	–	22.2	4.3	9.2	6.9
ENE	6.1	–	–	–	0.8	–
East	69.7	7.9	5.6	31.9	17.6	18.1
ESE	–	–	–	–	–	–
SE	–	–	–	–	10.1	4.5
SSE	–	–	–	–	–	–
South	–	1.6	–	2.1	1.7	2.1
SSW	–	–	–	–	0.8	0.3
SW	–	–	–	2.1	5.9	6.3
WSW	–	–	–	–	–	–
West	3.0	77.8	44.4	25.5	31.9	30.5
WNW	–	–	–	–	–	0.6
NW	–	–	–	2.1	5.5	5.4
NNW	–	–	–	–	0.4	–
Central	12.1	12.7	22.2	27.7	10.9	21.1
No. of howling sessions	33	63	18	47	142	266

responded to howling could not be determined. However, most howls were recorded from the east (17.6%, 18.1% for unidentified initiators and responders respectively) and west (31.9%, 30.5%). This supports the suggestion made above — that pack members initiated most howling from their own territory and responded most often when near the territorial boundary. When they were at the watering point both packs responded about twice as often as they instigated howling sessions (Table 5.2).

Initiators' and responders' physical actions after initiator howling are summarised in Table 5.3. Overall, the initiators were more active than the responders; they usually moved towards them, and often made contact and fought them, whereas the responders usually moved away. At the central shared resource (carcass at water), dingoes either stayed to maintain possession of the resource or departed hastily, depending on the availability of other natural food. Rodents were abundant there between March and May and both packs easily gave up the carcass when challenged. However, in July, when rodents were less available and there was also extra competition for the carcass from the group that had recently arrived from the desert, both the resident packs tried to keep the carcass (which fed several dingoes for several days) as long as possible.

Table 5.3

Initiators' and responders' physical actions (%) after howling by initiators.
I = initiator(s), R = responder(s)

Physical action	I %	R %
None	2.9	24.0
I and R move apart	5.9	28.0
I and R move closer	17.6	9.0
I and R join together, no antagonism	0	10.0
I and R join together, antagonism/fighting	23.5	13.0
Move to central shared resource (water/food)	5.9	10.0
Depart hurriedly from central shared resource	29.4	6.0
Maintain possession of central resource despite immediate presence of 'rival' group	14.7	0
No. of observations	34	100

Summary of howling by dingo packs in the Simpson Desert

Howling is mostly used by members of stable territorial packs (or subunits, see Chapter 6) and is most frequent when packs are using or defending essential resources — particularly oestrus and pregnant females, food and water. Howling is essentially used for long-distance communication and serves two purposes: attracting companions and repelling rivals. Dingoes distinguish these purposes by means of howl responses, sight, physical location, and pheromones to confirm the identity and perhaps the social status of both the initiating and responding howlers.

Plateau howls seem to provide non-specific information about the identity of the howler and are used for location per se; the message seems to be 'here I am, where are you?' Conversely, inflexion howls seem to provide specific information about the howler and a specific message. These howl types are mostly used by male individuals to attract or signal to companions and breeding mates.

Chorus howls clearly provide information on group size. Members howl together and use different types of howls. For large groups, some members may start with the same howl type but change the type of howl, varying the pitch and duration, so that rivals can tell how many animals are howling. The more dingoes that are howling, the more types of howls are used, and this is how numbers are 'counted'. Chorus howls probably also help to identify packs and are mostly used to repel rivals from essential resources.

Howling by dingoes living in loose associations in pastoral areas of central Australia

The same types of howls (plateau, inflexion, chorus) are used in this region and howling also has a highly seasonal trend, which is linked to the breeding cycle. However, compared to dingo packs in the Simpson Desert, dingoes here howl much less often (see Figure 5.3) and use short-range communication more often when they share water resources.

In arid pastoral areas there are probably no long-term stable packs because they are fractured by human activities (see Chapter 8), so that dingoes cannot maintain regular patterns of shared use of resources. Accordingly, one of the major functions of howling in these areas is to signal the intention of coming to water, to avoid unnecessary social strife. Dingoes give at least one if not several plateau howls (of low pitch and duration) or moans (described below) as they approach the water. For example, during one nocturnal observation session, two dingoes howled softly three times, but did not go to the bore to drink until the pair at water had drunk and departed. In another case, three dingoes came to water after much howling. No sooner had they arrived when other dingoes howled nearby, they looked about and quickly departed without drinking.

The most frequently recorded events were (1) single dingoes moaning continuously as they came to water to drink, and (2) dingoes already drinking moving off immediately another dingo howled nearby. The message of such vocalisations seems to be 'my turn to drink so please depart'.

As with the Simpson Desert packs, sighting in conjunction with howling also plays a major role in avoidance behaviour. Precedence at waterholes is sometimes decided by which dingo or group of dingoes sees the other first. If a dingo coming towards the water spots another dingo already drinking, it may stop and wait some distance away until that dingo leaves. Alternatively, if the approaching dingo is seen first, the dingo that is drinking may speedily depart.

Howling is also important for locating and regrouping of companions (or rivals) when hunting, as illustrated by the activities of three adult dingoes (2 males, 1 female) over 3 h in the Harts Ranges in central Australia. During the first hour, they remained more or less together, hunting for rabbits and lizards along a dry creek bed and adjacent areas. Then they left the creek system and soon became well separated, hunting by themselves. The female eventually caught a rabbit and ate it. One male was at least 1 km away from her, and the other male was about 0.5 km away in the opposite direction. About an hour after they had separated, one of these males gave a single plateau howl and then sat down near the crest of a hill, frequently looking about as if waiting for somebody. Five minutes later the female appeared and the male ran over and greeted her quite intensively. They then moved off to a nearby hill and sat down together, possibly waiting for the other male.

As with the Simpson Desert packs, more howling is heard in this region during the breeding season, particularly inflexion howls; they probably indicate the search for a breeding partner. Also, dingoes howl and answer howls irrespective of whether they are at or near water, which is unlike their behaviour in the non-breeding season. On one occasion, a male at water howled to a group of three dingoes on the other side of the creek, and they howled back. During February, when breeding groups begin to form, dingoes, particularly males, frequently run off in the direction of howls. On one such occasion a male obviously caught up with the howlers, judging by the sounds of fighting and squeals that suddenly came from that direction.

During the mating season dingoes respond to imitation howls made by humans in a singular manner. For example, a female came to water and remained for about 10 min, looking and sniffing about. Such behaviour does not occur at other times of the year when animals usually drink quickly and depart. When she eventually left, an imitation inflexion howl was given and she immediately turned and ran back towards the human howler (in an elevated hide). This performance was repeated eight times in as many minutes. The conclusion that this female was desperate for a mate seemed highly likely because imitation howling by the same observer at other times of the year usually made dingoes flee in the opposite direction.

There is also another, smaller peak in howling between October and December, which is mostly due to activities concerning the weaning and training of pups (Chapter 6). At this time pups are fully mobile and demand much food, and accompanying adults often kill large prey to satisfy them. Rather than drag the carcass to the pups, the adults frequently call the pups to the kill, where they give the characteristic excited, high-pitched, abrupt plateau howls.

Fortescue River region, north–west Australia

An index of howling (percentage of nights when dingoes howled) was highest in April–May, just prior to and during the mating season. This seasonal pattern of howling is consistent with howling by dingoes in all other regions of Australia. Although few data are available, the daily frequency of howling in this region is probably similar to that described above for dingoes in pastoral areas.

Barking and bark–howls

Dingoes can bark but it is sharper, more abrupt and more throaty than that of domestic dogs. Pure dingoes have not been recorded barking in the wild but some captive dingoes have learnt to yap and bark in company with their domestic inmates.

The bark-howl is an agitated cry, starting with one or several barks then followed usually by a plateau howl but sometimes an inflexion howl. In the wild, it has been heard only in situations of extreme alarm, to warn pups or companions of immediate danger. For example, early one morning, at 0200 h, a female was returning to her litter and unexpectedly discovered our observation site on top of a hill overlooking the valley where her den was. She immediately began to bark-howl. Her pups (about 3 months old) were camping near the den and immediately got up and disappeared down a nearby dry creek. Unlike other times when attendant adults howled, the pups answered once only, just after the female started, and they were not heard nor seen again while the female continuously bark-howled over the next hour.

On another occasion, a female was bark-howling and seemed reluctant to leave the area even though her male companion had just been shot and several shots were fired at her. When the shooter moved across a nearby dam wall for a better shot, a pup (previously unsighted) ran off and only when it was well away from the humans did the female stop bark-howling and also move away.

Dingo trappers have occasionally recorded bark-howling by one member (usually the female companion) of a pair of dingoes after the other has been leg-trapped. The free companion seems to be warning the arrival of the trapper. Similarly, at night in remote areas, dingoes sometimes bark-howl when they unexpectedly come across human campsites on their line of travel. Dingoes have also bark-howled when observers have lit cigarettes from atop of observation hides.

Moans

The moan is a soft-pitched howl rather like a plateau howl, that probably does not carry more than 100 m or so. It sometimes precedes dingoes coming in to water and has only been heard at such times. As indicated above, it is probably a signal to warn other dingoes at a waterhole that the newcomers wish to come and drink.

Snuffs

Dingoes occasionally snuff whenever they are startled by a (human) intruder that has unexpectedly come near them. When snuffing, dingoes rapidly and repeatedly expel air through their nose. They might be doing this to increase the intake of fresh air containing the scent of the intruder so they can quickly identify the intruder and decide whether to fight or flee. Snuffing and snuff-barking by wild dingoes have not been recorded, possibly because the sound does not carry to human observers. However, snuffing by captive dingoes was commonly recorded in Alice Springs whenever strangers visited their pens; it is interesting that Aboriginal people seemed to prompt more snuffing than Caucasian people did.

PHEROMONES AND OTHER TYPES OF COMMUNICATION

It is generally believed that dingoes and other canids ritually defaecate and urinate on objects at particular sites as part of their communication. It is also assumed that the pheromones (smells) associated with these deposits from anal glands, interdigital glands, caudal glands and possibly other glands convey to other dingoes a message, perhaps about the marker, or about the shared use of the area. Where do dingoes deposit smells? Do they really contain a message? If so, do other dingoes heed them?

Scent–posts: types, use and function

The objects (posts) upon which dingoes urinate and/or defaecate are many and varied. The most frequently recorded are grass tussocks, small bushes, fallen logs, fence-posts, rocks, and the faeces of other animals (e.g. buffalo, wombat, cattle). Scent-posts are established most frequently around shared resources such as water sources in arid areas, hunting grounds, and along trails and roads that are commonly shared by dingoes, particularly at intersections. The use of scent-posts increases in the breeding season. The following data illustrate how wild dingoes use scent-posts in arid and tropical Australia. Experiments with captive dingoes are then described, to investigate what messages pheromones may deliver.

At Alka Seltza in the Simpson Desert where the stable packs were operating, 163 faecal deposits were counted within about 100 m of the shared watering point (and carcass), and about 75% of these were on all of the available conspicuous objects (mostly grass tussocks). No faecal deposits were recorded further out, even though many objects there were suitable for marking.

During observations of the two packs and other dingoes (n = 23) over 12 months at Alka Seltza, dingoes urinated 69 times. Males urinated on conspicuous posts more often than females did (77% vs 9%). Males also urinated with a raised leg posture (RLU, see Plate 19) more frequently than females did (95% vs 1%). The males that used a squatting posture, similar to that of most females, were believed to be either young or of low social rank. After urination, males also raked the ground more frequently than females did (57% vs 15%).

The frequency of scent-post urination varied throughout the year. Along 111 km of the track leading to Alka Seltza, 10 fresh ground-rakings were counted in March and 24 in April, compared to only 1 in June and 3 in July. In the Fortescue River region in north-west Australia, levels of scent marking (RLU and ground-raking) increased over the 2–3 months before the mating period (May–June). Indices of RLU increased after January and peaked in April. Indices of ground-raking had two peaks (Feb and Oct) where the latter peak may have been related to the raising of pups. These data suggest that scent-post urination and ground-raking have a seasonal trend which is linked to the breeding season.

At Kapalga in the wet-dry tropics, 6,062 faeces were collected over 7 years from 58 transects more or less evenly spaced along a 134 km road network. The roads went through or terminated at all the major habitats on the 670 km^2 site. There were 14 transects at road junctions, 14 at road terminals at the junction of forest and floodplain habitats, and the remaining 30 were along the intervening stretches of roads. Relatively more faeces were deposited at the road terminals and junctions than on the roads connecting them (Table 5.4). For the terminal sites, 65.7% of faeces were on only four transects which were right next to a magpie goose breeding colony, an important prey species that provides 32.5% of the dingo's diet at Kapalga. Radio-tracking data indicated that two of these terminal roads also formed the territorial boundary between two dingo packs.

These data suggest that dingoes use scent-posts to indicate which hunting grounds they are sharing, and is supported by similar data from arid Australia. For example, on the edge of the Tanami Desert in the Northern Territory, a road about 20 km long joins two watering points in a red sand/spinifex habitat. This road is bisected by a narrow limestone belt which runs for many kilometres and contains large numbers of rabbits, the main prey of dingoes in this region. Along a 2 km stretch of road in the limestone country 56 faecal deposits were recorded, compared to none along the remaining 18 km of road in red sand habitat, and all these deposits were confined to about 20 m of the road. One explanation is that dingoes were using this part of the road to travel between waters, and were marking the turnoff point to the rabbit hunting grounds.

Other canids similarly deposit faeces at shared sites; for example, dholes regularly use communal latrines located at the intersection of two trails.

Table 5.4

The location of 6,062 faecal deposits (scent-posts) collected from road transects over 7 years (1980–86) at Kapalga in Kakadu National Park.

Dingoes deposit most faeces at sites where they are likely to meet — e.g. road junctions and road terminals leading to shared hunting grounds, rather than the stretches of road connecting them.

Location of road transect	No. of transects	No. of faecal deposits (mean ± sd)
Terminal	14	173.2 ± 190.5
Junction	14	135.7 ± 92.0
Stretch	30	57.9 ± 56.4

The role of conspicuous smells

Why do dingoes deposit scent on conspicuous objects and why do they rake the ground? They probably do it to draw as much attention as possible to their smells and whatever message they contain. If the scent is on an elevated object it will carry further, and a conspicuous scent-post also probably provides a visual guide for another dingo to home in on. Similarly, ground-raking may draw other dingoes' attention to the appearance of the disturbed earth or its fresh smell. When captive adult dingoes were allowed, for the first time in their lives, to investigate an undisturbed track through many trees and grass tussocks, the first male urinated and scratched around certain grass tussocks and one particular tree. There was no obvious reason why he chose them, but the next seven dingoes chose the same places for intensive investigation, and most also urinated and raked. Two months later the same dingoes and others, including females, concentrated on the same tussocks and tree.

On another occasion, several extracts that had been chemically isolated from anal gland secretions and urine (presumably pheromones) were placed on the ground in an area devoid of conspicuous objects to see how dingoes would react to them. The functions of different smells were not conclusively determined, but these experiments revealed more detail on the role of conspicuous scent-posts in advertising smells.

During these trials, dingoes sometimes walked virtually over the smell and did not become aware of it until they were downwind. Then they whirled about and sniffed their way back to it. Data collected on scent-posts in the wild suggests that conspicuousness depends on how many posts (objects) can be marked. For example, when two watering points were each being used by a similar number of dingoes, significantly more faecal deposits were recorded on the site that had fewer grass tussocks (potential scent-posts). Perhaps patchily distributed grass tussocks around a communal area, such as a watering point, are conspicuous because of their scarcity, so that a dingo feels obliged to investigate them, and in doing so, discovers that one of his colleagues was there before him, and accordingly urinates and so on.

In another experiment, captive dingoes were led along a track which included a bare patch of ground (about 10 x 4 m), which they ignored. Then imitation dingo rakings were made, and the same dingoes, along with others, were led past this site. The males stopped at the rakings, sniffed about, then kept going; but females ignored them. Then a grass tussock (one that these dingoes had previously ignored) was planted at the apex of the rakings, and the same group of dingoes was led past it. Again, all the females ignored the setup, but all the males stopped and sniffed at the tussock and one male made a half-hearted attempt to urinate.

One of the functions of canids' scent marking may be to synchronise reproduction between pairs. Studies of canids other than dingoes indicate that gender and endocrine information can be conveyed by urine, and testosterone and oestrogen both increase the frequency of urination and heighten their interest in odours of the opposite sex.

A dogger's version of pheromones

Despite the years of costly research by eminent scientists, the functional significance of pheromones in canids, particularly anal glands, is still not well understood. There is a dogger's anecdote about why dogs sniff the 'backsides' of other dogs that may seem, to some people, as convincing an explanation as scientific results. The yarn goes something like this. One day, in the beginning, all the dogs in the world met for an important meeting. As they filed into the auditorium they hung their 'backsides' on the pegs provided, just as men used to hang up their hats in more formal times. During the meeting a great conflagration broke out, and in the melee the dogs grabbed whatever 'backside' they could in their rush to safety. And to this very day, dogs are still checking out 'backsides', looking for their own.

CHAPTER 6

SOCIAL DYNAMICS

This chapter first describes the structure of a dingo pack in captivity to suggest how wild packs are formed and maintained. Then it presents field evidence to show how free-living packs operate in contrasting habitats throughout Australia and in Asia. It appears that fully operative dingo packs, akin to their wolf forebears, are only seen these days in remote areas where dingoes are seldom disturbed by pastoralism or human control activities.

THE SOCIAL DYNAMICS OF A DINGO PACK IN CAPTIVITY: CURLY'S MOB

In a study made in Alice Springs, a pair of adult dingoes, Curly and Toots, were kept for three breeding seasons in a large open-air enclosure containing water, trees, shrubs, grassy areas and a large pile of rocks that provided breeding dens and sheltered them from inclement weather and aggressive inmates. They and their progeny were allowed to reproduce and otherwise interact, with minimal human interference. The behavioural data that was gathered indicated how packs are established, organised and maintained, and how breeding is suppressed by wild dingoes.

Births and deaths

Toots, a female 3.5 years old when the study began, successfully produced litters in each of the three breeding seasons (5, 7 and 4 pups respectively), all sired by Curly, initially 6.4 years old (Figure 6.1). Each of her daughters came into oestrus at about 11 months, all became pregnant, but none of their pups survived.

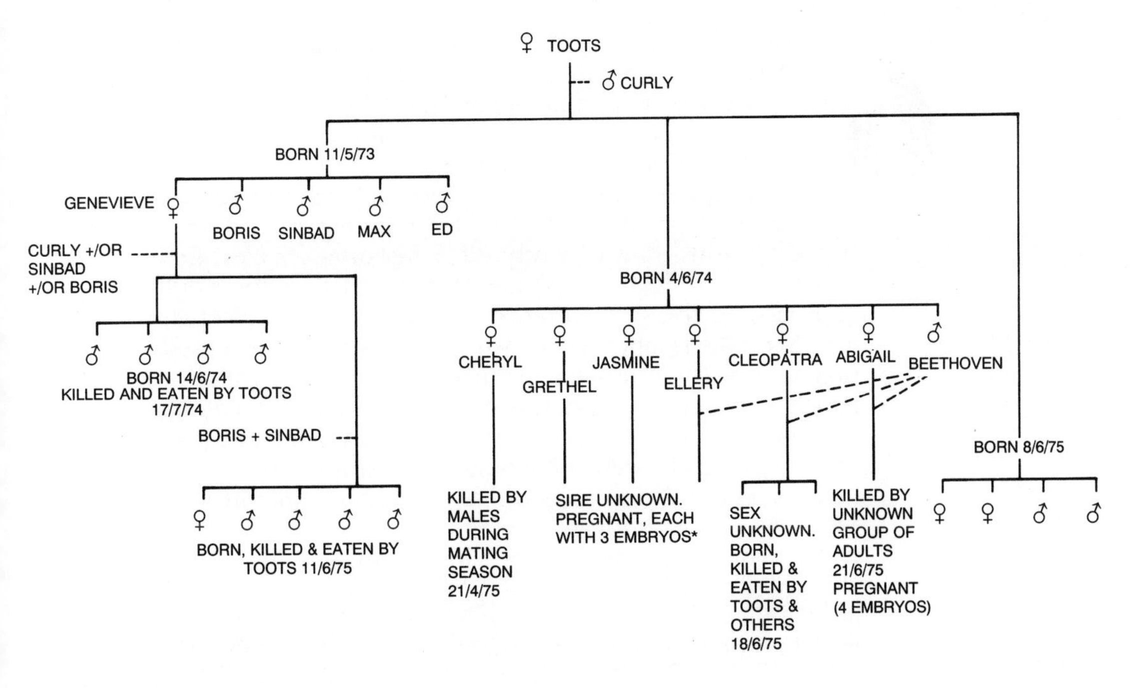

Figure 6.1
The genealogy, births and deaths of a colony of captive dingoes (Curly's Mob) in Alice Springs over 3 years (from Corbett, 1988).
--- known sires
* experiment terminated before whelping

Genevieve was the only daughter in Toots' first litter, and after mating with her father (Curly) and two of her brothers (Sinbad and Boris), she whelped 4 pups. Two days later, Toots, who had whelped 10 days previously, shifted her litter of 7 pups into Genevieve's den. Both mothers independently attended this den containing both litters, now mixed, and attempted to suckle all the pups. However, over the next 4 weeks, Genevieve's younger and smaller pups became progressively weaker and were probably starving as a result of competition for nipples with Toots' older and larger pups. Then, 7 days after Genevieve's pups started moving around outside the den, Toots, with assistance from her pups, killed and ate all Genevieve's litter. At the time, Genevieve looked very apprehensive, but neither she nor the adult males attempted to stop the infanticide.

The next year there were 8 sexually mature females. During the mating season in April one oestrus female (Cheryl) was killed by a sexually excited male group comprising her sibling brother (Beethoven) and 4 older brothers (Boris, Sinbad, Max and Ed). Toots was again the first female to whelp, and 3 days later Genevieve gave birth to 5 pups including a stillborn runt. That same day Toots transferred Genevieve's pups to her den, then killed and ate them. Again there was no interference from Genevieve nor from the males, and both females subsequently suckled Toots' 4 pups.

One week later, Cleopatra whelped 3 pups and all were immediately killed by Toots, then eaten by her and several of her adult sons. Subsequently all 3 lactating females alternately suckled Toots' pups and threatened all other adults approaching the den entrance, except for Curly. Cleopatra also persistently attacked and harassed Abigail, her heavily pregnant sister. Beethoven, the brother who mated with both females, also assisted in harassing Abigail. She was eventually killed 3 days after the birth and death of Cleopatra's pups. The nature and extent of wounding indicated that Abigail had been killed by a group of adults that probably included Cleopatra and Beethoven.

The remaining three females (Grethel, Jasmine, Ellery) were also pregnant and were also frequently attacked by Cleopatra and Beethoven, and occasionally by other males except Curly. Whether they or their pups would also have been killed was not determined as the experiment was terminated a week after the death of Abigail.

Formation of the pack

The pack was essentially formed when Toots' first litter matured, and comprised 2 adults (1 male, 1 female), 4 subadults (3 males, 1 female) and Toots' litter of 7 pups (total 13). After the breeding season there was always one subadult male (the outcast) who did not interact with pack members. Except for defending himself from their often combined harassment, he lived in virtual isolation in the rock refuge and only emerged to drink and snatch food. Thus there were 6–7 dingoes participating in the social ranking.

In the following breeding season, the pack comprised 6 adults (4 males, 2 females), 2–3 subadults (1 male, 2 females) and Toots' litter of 4 pups. Of these only the adults and the subadult male (Beethoven) interacted in the social hierarchy. The three subadult females were shown tolerance by pack members probably because they were in oestrus or were lactating and suckling Toots' pups. After the breeding season, pack size (12 members) and the number of dingoes interacting in the social order (7 participants) were both similar to that of the previous year. The number of other dingoes living separately from the pack had correspondingly increased; there was one adult male outcast (not always the same individual) and there were three subadult females.

Overall trends in social interactions

Over the first two years of the study the number of interactions per dingo generally increased with the number of dingoes, but this trend was reversed in the next year despite the presence of more adults (12–14). This reversal occurred because not all dingoes participated in the social order, and because variation in the frequency of particular behaviours was related to the sex rather than the number of participating dingoes, as indicated below.

Aggression

Overall, aggressive interactions (54% of all social activities) outnumbered active submission and other behaviours (Figure 6.2a). The trend for active submission was inverse to that of aggression, which indicates that active submission probably signifies rank stability in the pack. As Toots' first litter of 4 males and 1 female matured, the monthly proportion of aggressive interactions steadily increased, but that trend was reversed the following year when Toots' second litter of 6 females and 1 male became part of the pack.

Aggression was more frequent between males (mean of 1.4 interactions/hour/dingo) than between females (0.3) or between the sexes (0.2) (Figure 6.2b). For males, most aggression occurred during the non-breeding months, and the overall level peaked in the second year of study (5 males) then steadily declined the following year, even though 6 males were present (Figure 6.2b). That trend suggests that male aggression was associated with pack formation after Toots' first litter matured and after the pack stabilised; aggression then declined.

Males received an average 12 wounds/dingo/month, which was twice as many wounds as females received. About half of all wounds were on the head and shoulder regions, which suggests that most were associated with threat behaviour by two dingoes. Wounds on other body regions usually occurred when several dingoes combined to attack another. The two subadult females (Cheryl, Abigail) who were killed by a group of dingoes received deep lacerations in the upper hindquarters and back with some puncture wounds penetrating the kidneys. Males received most wounds in April and May, the months when all copulations occurred. This indicates that aggression was most intense then, even though aggressive interactions were relatively few. In contrast, most females received most wounds in June when some were lactating. There were significant differences in the number of wounds received by individuals in both male and female ranks (range for males, 2–24/month; females, 4–13/month). The alpha pair, Curly and Toots, received fewer wounds than other pack members did.

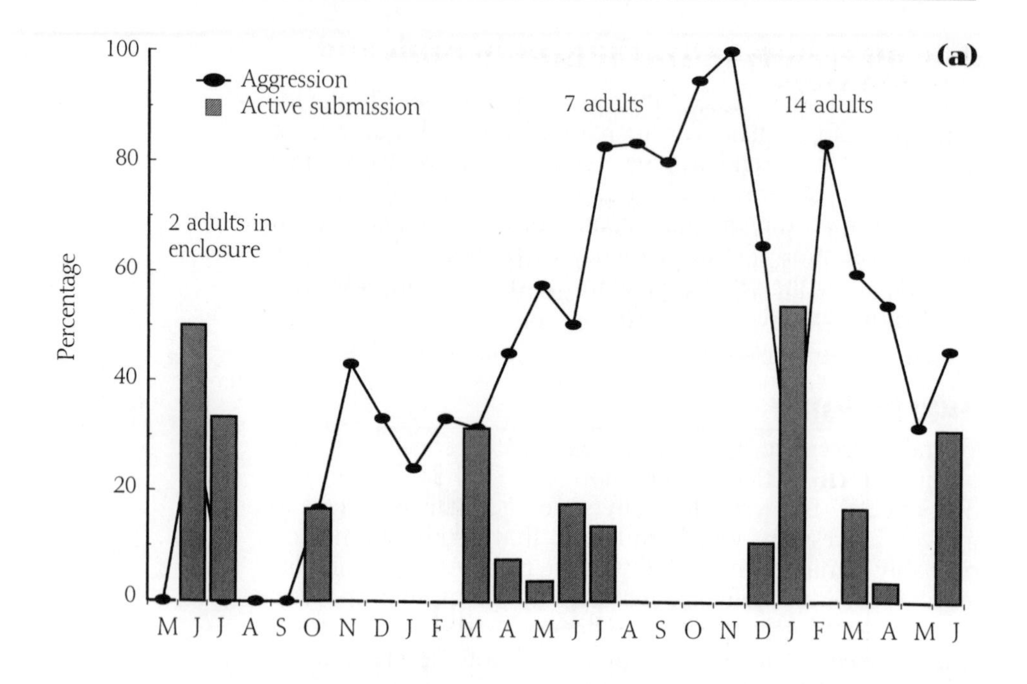

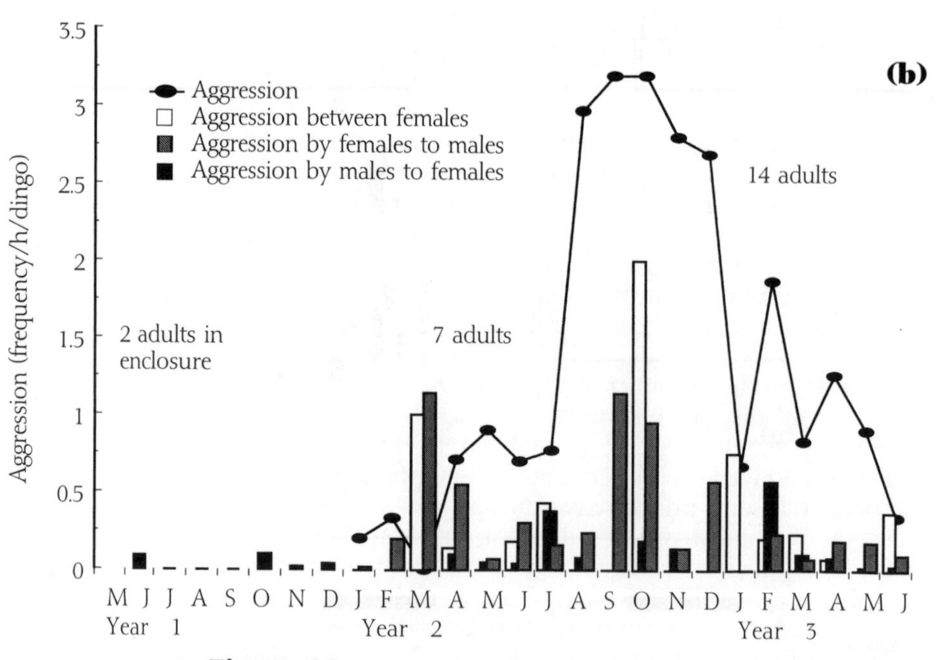

Figure 6.2

Aggression in a captive colony of dingoes (Curly's Mob) over 3 years (from Corbett, 1988): (a) monthly percentages of adults' aggressive behaviour and active submission; (b) monthly frequency of aggression within and between sexes.

The male hierarchy: changes in rank and related stress

Males were clearly assigned a specific rank. Apart from Curly, who always occupied rank 1, all other males fluctuated, irregularly, between high and low ranking; for example, Boris changed rank 35 times but most often occupied rank 2 (Figure 6.3c). Overall there were 135 rank changes (mean 8/month) and significantly more occurred during the breeding months (Feb–July). Relatively more changes occurred in the third breeding season when the enclosure contained 6 males (Figure 6.3g). That peak significantly correlated with

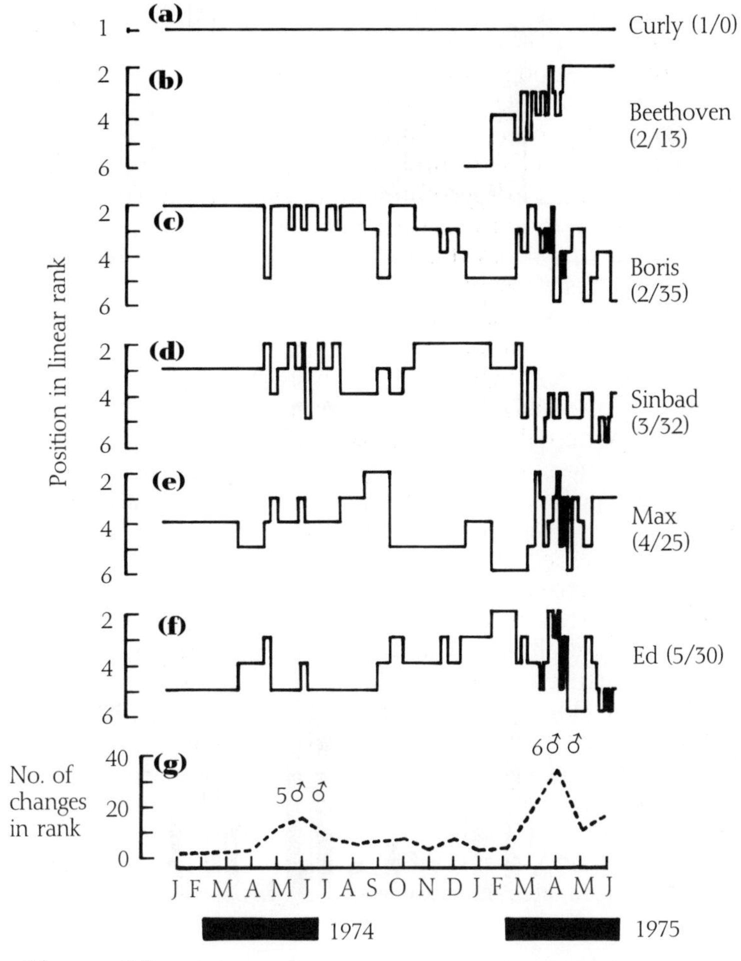

Figure 6.3

Fluctuations in social rank of male dingoes in a captive colony (Curly's Mob) over 3 years (from Corbett, 1988). (a)–(f) Positions and their approximate duration for 6 adults. Figures in parenthesis give rank mode and number of rank changes for each male. (g) Monthly changes in rank for all males. Solid bars denote breeding seasons.

the incidence of wounding but not with the frequency of aggressive encounters. These data indicate that there was a male hierarchy which, apart from the alpha male, was least stable during the breeding season. Although relatively fewer aggressive interactions occurred then, they were more intense and resulted in serious fighting.

About two-thirds (67%) of all aggressive interactions were instigated by high-ranking males towards low-ranking males. Of these, most (42%) were instigated by the alpha (rank 1) and equally directed towards high-ranking (ranks 2–3) and low-ranking (ranks 4–6) males. There were only four aggressive acts directed at the alpha and all were unsuccessful. Aggressive interactions between high-ranking (12%) and low-ranking males (15%) were similar.

Curly, the alpha male, not only instigated significantly more aggressive interactions than other males did (mean of 2.4 interactions/hour/month compared with 0.8 and 0.7 for high-ranking and low-ranking dingoes respectively), but there were peaks in aggression about every 3 months which generally corresponded with troughs in aggression instigated by low-ranking males (Figure 6.4). Curly, therefore, was not only continually 'on his guard', as indicated by his constant high level of aggression, but it also appeared necessary for him to regularly reinforce his alpha status.

The constant, high levels of instigating and receiving aggression probably has physiological costs, such as stress, which may be reflected by measuring plasma cortisol levels in blood. These levels were highest in low-ranking males (mean 5.1 μg/100 ml) in the early breeding months. This matched their relatively nervous behaviour in the presence of other males especially since they were the recipients of most aggressive behaviour. Cortisol levels in individual males tended to increase threefold when they were low-ranking and vice versa, but all values were less than 9 μg/100 ml — that is, they stayed below the level that is thought to indicate stress in domestic dogs. Surprisingly, the alpha male (Curly) also registered high cortisol values which, along with his high, constant level of aggressive interactions, suggests that maintaining the alpha position may also be stressful and may partially explain why Curly did not persist in suppressing the breeding activities of subordinate dingoes after he had mated with the alpha and beta females.

Female hierarchy and male–female hierarchy

Females also had a clearly defined hierarchy, but unlike males they made few changes in rank. Toots was always dominant to Genevieve and both were dominant to the other six females, who were almost always low-ranking with little

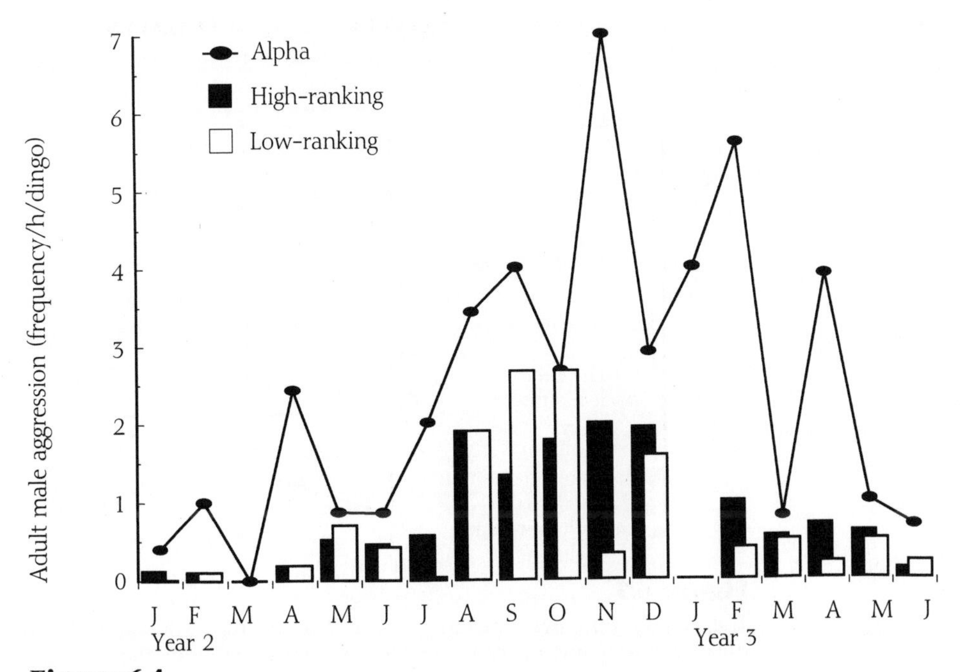

Figure 6.4
The monthly frequency of aggression initiated by the alpha male, and means for high-ranking and low-ranking adult males in a colony of captive dingoes (Curly's Mob) over 3 years (from Corbett, 1988).

difference between them. When in oestrus or suckling Toots' pups, they briefly occupied a higher rank, similar to Genevieve's. However, this elevated status was achieved because attendant males helped them win disputes with other low-ranking females. The alpha and beta females tended to have double the plasma cortisol levels than those of females of lower rank (mean 2.7 versus 1.4 μg/100 ml), and values for all females were highest in the breeding season.

The hierarchy for both sexes was clearcut, although it varied between the breeding and non-breeding seasons. Curly was always dominant over all other dingoes, Toots was always dominant over all dingoes except Curly. Apart from the outcast, the other males were usually dominant over the other females. During the breeding season, oestrus and lactating females were shown tolerance which enabled them to achieve, briefly, a higher status. Figure 6.5 presents this dingo hierarchy.

THE SOCIAL DYNAMICS OF WILD DINGOES IN AUSTRALIA

Given the wide variety of habitats, prey species, climatic conditions and levels of human exploitation encountered by dingoes throughout Australia, it is not surprising that the social

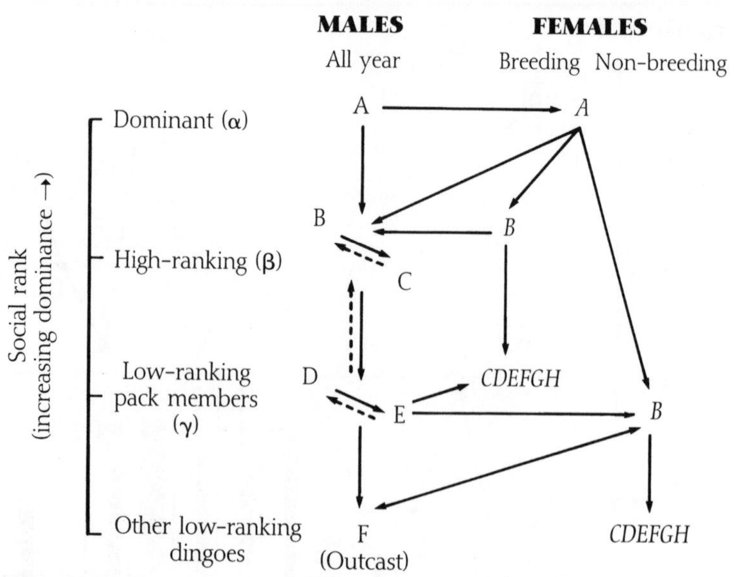

Figure 6.5

The dingo's dominance hierarchy derived from dingoes in captivity in Alice Springs (Curly's Mob) over 3 years. Females are indicated by italics. A dominates B and all others underneath; B dominates C, etc. Dashed arrows connect males most likely to change status. Females *C-H* rotate to a high-ranking status when in oestrus or lactating; at this time males display tolerance behaviour, not submission. Female A is dominant whether breeding or non-breeding.

organisation of dingoes is so variable. The following cases illustrate the variation in the group size and social behaviours of dingo populations living in different regions of Australia.

Stable large packs in the Simpson Desert

The distinguishing feature here is the persistence of large, stable packs of up to 10 members (Table 6.1) that consist of young and old animals of both sexes similar to Curly's Mob described above. Packs are territorial and adjacent territories usually radiate from isolated and shared watering points (see Plate 15). Typical howling interactions between two such packs were described in Chapter 5 for the West and East packs at Alka Seltza.

Copulation was observed once between the alpha pair and, remarkably, there was absolutely no interference by other pack members such as occurred in the unruly melee exhibited by captive dingoes and other canids. This indicates that the social hierarchy of each pack is very stable and that members of these desert packs stay together for long periods.

Although a pack contains two or more mature females, successful breeding is restricted to one female (probably the

Table 6.1

Group sizes (% of observations) of adult dingoes in different regions of Australia.

Although individuals and small groups are most commonly recorded, the assumption that dingoes are essentially 'solitary' is not accurate because it does not acknowledge the associations between apparent loners indicated by radio-tracking studies and by almost continuous observations made over several months at one site. n = 287, 400, 3,624 and 1,000 observations for the Simpson Desert, Kakadu National Park, Fortescue River and central Australia respectively; + indicates the size of groups recorded in the eastern highlands (no data on frequency). The largest group of dingoes (23 mature and juvenile animals of both sexes) ever recorded, in the Fortescue River region, was associated with hunting and feeding on large macropodid prey.

	Group size							
	1	2	3	4	5	6	7	≥8
Simpson Desert								
Total	48	18	13	9	3	3	5	1
West Pack	34	22	19	9	5	2	9	0
East Pack	26	18	21	16	5	5	9	0
Loners	95	5	0	0	0	0	0	0
Desert Group	43	0	0	14	14	29	0	0
Unknown	52	22	12	9	2	2	2	1
Kakadu NP	66	19	9	1	2	1	1	1
Fortescue River	25	21	17	13	7	6	5	6
Central Australia	73	16	5	3	1	1	1	0
Eastern highlands	+	+	+			+		+

alpha female), so that each pack raises only one litter annually. This pattern is similar to wolves, African hunting dogs and dholes. Precisely what inhibits successful reproduction in subordinate dingo females is not known. Since some of these females appeared to become pregnant, the inhibiting factor is probably dominant female infanticide, similar to that which occurred in Curly's Mob (see Plate 18).

Non-pack animals (loners) also share the waters and carcass food. They operate in the corridor between the territories of resident packs, and are most active during daylight hours when pack members are usually resting. Most loners are young animals less than 2 years old, and are probably offspring of the alpha pair born in a previous year. Such young females tend to remain in their natal areas whereas males disperse.

During years when widespread, substantial rains persist, many packs spread out into the desert and pack members probably become dissociated on account of the abundant, accessible food (mostly small mammals) that also increases with the rain. Later on, when ephemeral waters decline but rodent populations have peaked, dingoes seem able to survive on the water contained in prey without drinking from watering points. Eventually, most dingoes need to return to the edge of the desert where water is more plentiful and relatively permanent, as at Alka Seltza. It is then that competition and interactions with local resident packs is most intense, as indicated in Chapter 5.

Because waterholes are usually shared and thus the focal point for interactions between adjacent (territorial) packs, dingoes are naturally wary when they approach them. In doing so, they perform a ritual display of 'cautionary bravado'. For example, soon after one pack had departed, the new arrivals came to within about 100 m of the water, then waited to make sure no other dingoes were around. After several minutes the leader (alpha), trotted towards the water to reconnoitre, tail curled over his back, hackles raised, followed by the other members of the pack in similar postures; they were in a compact group and making sure they were not alone. Near the water's edge they abandoned caution and, full of bravado, stormed around in a tight unit, dust rising, just like a bunch of Hell's Angels hitting town on a Saturday night (see Plate 17).

Stable large packs in Kakadu National Park

Dingoes here may be solitary, in pairs or in groups of 3–8 (Table 6.1). The main activity periods are dawn and dusk and the first half of the night. Howling and scent-marking seem important for maintaining territories and gaining access to shared hunting grounds.

The largest groups (territorial packs) are seen most frequently between November and February and least often in June; pairs were most often seen between March and May. Radio-tracking studies indicate that variation in pack size and cohesiveness is usually related to the wet months (Dec–Feb) and dry months (May–Nov). These periods determine where dingoes hunt (upland forest or floodplain habitats); what they hunt (wallabies or rats and geese); and how they hunt (group or solitary hunting). Dingoes can reach the floodplains only during the dry season to hunt geese and rats, the major prey species in this region (see Chapters 7 and 9).

Members of territorial packs hunt communally for buffalo and large macropodids, but when they hunt rats and grasshoppers on the open dry floodplains they spread out and hunt independently. Nevertheless pack cohesion is still maintained by sight in the day and by howling at night in what can be described as a 'mobile' or 'fluid' territory. For example, during an early morning observation on the flood-plains where even the human eye can 'see forever', 13 dingoes remained in (binocular) view over an approximately 10 km^2 area, and for 4 hours no individuals came in contact with each other; in fact, they seemed to avoid each other. Yet, from radio-tracking many of the same animals at other times it was known that they belonged to two discrete packs, and that in other forest locations opposing members invariably had aggressive interactions.

Only one litter per pack is usually recruited into the population. At Kapalga (670 km^2) there were 5 packs (6–8 members) and about 25 pups were recruited each year, which kept the total dingo population fairly constant at about 60 adults for over 10 years. Most litters are whelped between June and August, although it is known that breeding occurs in every other month. Such breeding may be associated with subordinate females (who temporarily leave the pack) or loner (non-pack) animals. The largest recorded litter of 10 pups may have been a combination of two litters (similar to Toots' and Genevieve's), but the fate of these pups was not determined.

Stable large packs in the Fortescue River region

From aerial radio-tracking studies made over 9 years, 25% of dingoes were recorded alone, 21% were in pairs and 54% were in groups of 3–23; the mean group size was 2.1 (Table 6.1). However, as studies in other regions found, the relationships between radio-collared dingoes indicate that most are members of discrete packs (mean monthly pack size 3–12 members) that occupy essentially non-overlapping living areas.

Packs comprise a mated pair and their young from previous years, similar to Curly's Mob, and although pack members interact in communal activities such as foraging, the full pack is rarely together at any one time. A social hierarchy also similar to that of Curly's Mob, probably exists in the Fortescue packs, since particular dingoes (presumed alphas) were observed consistently leading the pack when travelling and feeding at carcasses before others, and low order (subordinate) dingoes were observed feeding at carcasses only after all other pack members had finished.

The mean group size of packs varies little over the year: from 2.4 in the mating season (May-June) to 1.8 in the nursing period (July-Sep), to 2.3 in the non-breeding season. The relatively higher group size during the pre-breeding season, along with increased scent-marking and howling, indicates a high level of pack cohesion then; this is consistent with data from all other regions in Australia where dingoes have been studied.

In this region, pack size does not correlate with territory size since, for example, the largest pack ever recorded (mean 12 members) occupied an area similar in size to that of the smallest pack (mean 3 members). Instead, pack size seems related to the availability of various resources. For example, the territory of the smallest pack also had the lowest density of main food (kangaroos), and the fewest watering points, which were also less widely distributed.

The territorial boundaries of dingo packs remain quite stable from year to year, which indicates strong site attachment, and it seems there are very few 'physical' interpack encounters. In a study of 5 packs over 4 years, only 5 encounters between neighbouring packs were recorded. All involved a single dingo from one pack interacting with at least two members of the other pack; one encounter seemed friendly, one involved avoidance and one involved fighting.

Raising pups in the Fortescue River region

In the Fortescue River region few dingoes other than the mother are seen much closer than 1.5 km of the den during the first 3 weeks of the nursing period. The pups are first seen outside the natal den when they are about 21 d old. The natal den, or a secondary den (0.5–1.4 km away) is abandoned when the pups are about 8 weeks old. Then the pups occupy a succession of rendezvous sites, usually rocky ledges about 3 km apart, for various periods (mean 3 weeks, range 6 days–6 weeks) until they are fully weaned. The pups usually roam about at these sites by themselves, but are accompanied by adults on longer jaunts of more than 200–300 m.

Alloparental (helping) behaviour regularly occurs; pack members (males and females) help the mother to bring food to the pups. On one occasion, a mother gorged on a kill, returned to the den and regurgitated food to the pups. Soon after an adult male arrived at the den and regurgitated food to the pups and the mother. Such behaviour is similar to that of African hunting dogs and dholes where males also 'guard' the pups when other pack members are away hunting; and African hunting dog males are also capable of raising the pups when the mother dies.

As the pups grow older, pieces of prey (e.g. kangaroo legs) are brought to the pups at the rendezvous site. Later on (at 9–14 weeks) pups are taken to feed on nearby kills before returning to a rendezvous site. From then on, pups accompany adults (mean 9–24 weeks). Most pups are weaned by 16 weeks, although they remain in the company of adults and share in pack activities.

Lone dingoes in the Fortescue River region

Very few lone dingoes were observed during the 9-year Fortescue study; 11 dingoes became dissociated from their packs (6 females, 5 males, 7 yearlings, 4 adults) and although their large ranges overlapped the mosaic of pack territories, most seemed to avoid encounters with packs. One male was regularly recorded in the presence of one pack and may have been attempting to form ties with it, but he was not successful. Lone dingoes are probably in various stages of dispersal; they roam over large areas looking for mates or for vacant areas to settle in.

Unstable associations in arid and semi-arid pastoral regions of Australia

Since the pastoral industry occupies large areas of Australia, the descriptions of dingo sociality below are probably typical for most dingoes in Australia.

Of 1,000 dingoes recorded by a research team over 8 years in central Australia, 73% were seen by themselves, 16% were in pairs and 11% were in groups of 3–7 (Table 6.1). Because dingoes are essentially inactive during the day, people mistakenly conclude that dingoes are essentially solitary animals. However, the 'solitary' dingoes they see often have unseen companions resting nearby. Many bushmen know that one of the best ways to view a dingo close up is for two vehicles to drive through dingo country; the first driver may see nothing or perhaps one dingo, but the second driver invariably sees dingoes that are disturbed by the first vehicle and come out on the road to sniff the vehicle tracks. Radio-tracking studies in several regions of central Australia have confirmed that apparently 'solitary' animals are loosely bonded in small amicable groups ('tribes') that share a single living area but tend to hunt separately. Unlike the pack territories described above, the living areas of adjacent 'tribes' overlap considerably and are not overtly defended throughout the year. The three dingoes hunting rabbits, described in Chapter 5, exemplify such a 'tribe'.

Activities at a den site in the Harts Ranges, central Australia

The following example is typical of how dingoes cooperate to rear pups; it involves observations of a litter of pups and attendant adults in the Harts Ranges in central Australia where the main prey is rabbit. The den was an excavated rabbit hole in a large warren, and was first observed in early September when the five pups were about 10 weeks old. They had recently stopped suckling and were quite active, spending most of the day in a nearby shady dry creek bed, and camping in the den at night.

Greetings between family members and helpers

The mother (Mum) returned from a foray and one pup came bouncing toward her, soon followed by the others and all greeted her affectionately with much muzzle licking and tail wagging (active submission). For the next 2 h until sunset Mum rested and the pups either played with each other, which involved much hiding, stalking, rushing and knocking over, or they stalked one of the many rabbits nearby.

About 60 rabbits lived within 100 m of the den site and they seemed quite unperturbed about the presence of the female and the pups. They seemed to know that the pups were only amateur hunters, because when the pups came stalking they waited until almost the last moment before popping down a hole.

As the sun was setting, Mum was still resting but looking about more frequently as if she was expecting somebody. Sure enough, an adult ginger male dingo (Dad) came trotting in from the west along the valley and, even more surprising, he had a collar around his neck which scientists had fitted the previous year. As he approached the den site, he urinated twice on bushes and ground-raked. When he saw a pup coming towards him he curled his tail and held his head high with ears forward and strutted along in a most aloof manner. He seemed to ignore the pup, who seemed hesitant about coming any closer, and kept on strutting along and urinated and ground-raked again, then stood looking around. By this time another pup had come to meet him, but he did not greet the pups, nor did they greet him, although they obviously wanted to.

Then Mum, about 30 m away, stood up, looked at Dad, flattened her ears and trotted towards him. Dad's tail straightened out and wagged from side to side. Mum approached, crouching slightly with her ears flattened sideways and her tail about 45° down and wagging from the hips, in a typical display of active submission. When they met, Mum leaned against and underneath Dad, licking and gently biting his muzzle. Dad stood there and accepted this but now his tail was curled over his back. When Mum greeted Dad the two pups ran across and greeted both adults in a similar manner to Mum's, their ears flattened sideways and tails wagging

vigorously. Another two pups appeared and similarly greeted the adults.

After accepting the greetings for about 2 minutes, Dad strutted off followed by Mum, still licking and crouching, and five frolicking pups; the whole scene was one of the most fantastic sights one can imagine. Both adults then trotted off to the west gradually gaining speed, and about 100 m from the den, the pups stopped following and the adults disappeared behind a hill. Mum was leading, both adults had their tails straight out and they had purpose in their stride as they headed out to hunt rabbits. The pups slowly made their way back, playing all the time and eventually settled down in the creek bed. The pups appeared to know they had to stay behind as there was no obvious signal from the adults.

Feeding the pups

At 2211 h, Mum returned with a rabbit for the pups, and there was much growling and snapping amongst them and a 'tug of war' with the rabbit. Soon afterwards Mum departed again and did not return until 0200 h the next morning when there was also a short burst of howling by one adult and one pup. For the rest of the morning and the next afternoon, no adults were seen and the pups either wandered aimlessly around, or rested, or played with each other, or tried stalking rabbits.

At 1745 h, about half an hour before sunset, three adults came up the creek from the east: the collared male (Dad), another ginger male (Dave) and a white female (Mabel). Two of the pups greeted Dad with active submission, and he seemed more tolerant of them than thrilled to see them. The pups did not greet the other adults, but approached them with their tails between their legs. These adults just stood and sniffed noses with one of the pups once. After several minutes the adults disappeared over the hill, leaving the pups behind. Again there was no visible signal from the adults telling the pups to stay. However, about a minute later a low, fairly short howl came from one of the adults, and it was answered 3–4 times by the pups. Two pups also ran several metres in the direction of the howl.

Similar activities occurred over the next 2 weeks. There seemed to be no fixed time when adults went hunting and, naturally, there was no fixed time for returning with food, which happened 2–3 times a night. Mabel was a young virgin and, despite her white coat colour, was probably one of the previous year's litter. She was a helper, and it seems a sensible system, since one female feeding five pups and being able to carry only one rabbit at a time would otherwise spend a great deal of energy. On most of the dusk hunting trips the adults first went 1.5 km to drink at the nearest watering point (windmill and trough). For most of the daylight hours they rested in a dry creek about 250 m from the den.

Watering the pups

As the weather was hot and the pups had stopped suckling, the observers wondered at first how the weaned pups obtained water. Soon after dark Mum returned directly from the bore and regurgitated water to the pups. During this process the pups were quiet and licked at her mouth, unlike the times when she brought food — they were boisterous then, and often fell over each other.

Teaching the pups

During October there was more howling by adults and answering by pups. The pups were also going further afield for longer periods, sometimes by themselves. At about the same time the rabbits near the den became very jumpy and most disappeared down a hole at the mere sight of a dingo. It was obvious that the pups had become more advanced in the art of hunting rabbits. Instead of rushing blindly at the rabbits as previously, they were now deliberately stalking them, and rushed over the last few metres only when they had some chance of success. The pups were now much bigger, had lost their floppy, puppy style of running, and were now quite fast and determined.

Mabel was observed coaching the pups in the art of stalking. She would run for several metres then stop and wait for the pups to catch up, then she would chase after rabbits again. The pups mimicked her movements, stopping when she stopped and running when she ran. It seemed that the local rabbits had been saved for the pups.

Weaning the pups

In November the pups were rarely seen at the den site, but adults and pups still did much howling; apparently the pups were learning the basic rules of communication. From December to February, when the weather was quite hot, the pups were based at the watering point, and appeared to have been totally weaned from their parents and helpers, although they occasionally met each other at the watering point. At this stage a mineral exploration camp was set up near the watering point, and after dingoes repeatedly raided its stores and ripped beds, three pups were shot and poisoned. Another of the pups was shot by the station lessee. Such events are not uncommon and indicate how dingo society can be fractured by human activities.

In any event, most pups are abandoned by their parents and thrown on their own resources when about 6 months old. Subsequently, however, they are frequently seen in the company of adult males, following their every move, which suggests that after leaving their parents young dingoes refine their hunting and other behaviours by imitating their elders.

Maternal care in the Simpson Desert

The care of pups also varies depending on the type and abundance of food. For example, on the gibber plains at the western edge of the Simpson Desert, rabbits are much less abundant than in the Harts Ranges, and the supply of other prey such as rodents and marsupial mice is unpredictable and spasmodic. In this region grasshoppers often form a major component of the pups' diet. They include large flightless species weighing up to 5 g.

As soon as the pups are mobile (at about 4 weeks) the mother leads them about the countryside following concentrations of grasshoppers which are caught by both the mother and the pups. When the pups stop suckling (about 6 weeks) and their food requirements are much greater, the mother splits the litter into smaller units and settles them in separate places, often several kilometres apart, so that they can feed on the local concentrations of grasshoppers more efficiently.

On one occasion, not unexpectedly, two mothers settled sub-units of their litters in the same area. But, unexpectedly, the two litters intermingled and both mothers combined efforts to feed them all. When the local supply of grasshoppers declined, the mothers departed in separate directions, but each with a mixture of both litters; this was deduced from their coat colours (one litter was black-and-tan, the other ginger).

Semi-stable small groups in east and south-east Australia

Despite extensive studies in this region, very few data have been published on the social behaviour of dingoes, partly because there are many hybrids in wild populations here (see Chapter 10) that are difficult to identify in the field and whose behaviour may differ from that of pure dingoes.

The available data suggests that in this region most wild canids (dingoes, feral dogs and their hybrids) are organised into small groups of 1–3, although larger groups of up to 6–9 animals have been seen or registered by radio-tracking studies (Table 6.1). The incidence of small groups may be partly due to human control activities which fracture large groups of related individuals, and partly because small groups survive better when mostly small and medium-sized prey are available. For example, near Georges Creek Nature Reserve in north-east New South Wales, the dingo social structure changed from small to large units when the source of available macropodids changed; presumably, larger groups of dingoes were necessary to kill large grey kangaroos when they became more prevalent than wallabies.

Similarly, at Nadgee Nature Reserve, dingoes are usually seen alone or in pairs, but may combine into larger groups to hunt large macropodids. In Kosciusko National Park a group of 8 adult dingoes was once seen near the carcass of a 6-month-old horse, and apparently they had killed and eaten it. Another radio-tracking study made in Kosciusko National Park concluded that up to 9 animals shared one living area, but only up to 3 of them were together at any one time.

Comparisons between regions in Australia

There are two broad levels of comparison: between two or more 'wilderness' regions, and between 'wilderness' and pastoral regions. The main difference in the operation of dingo societies in 'wilderness' areas, such as in the Simpson Desert, Kakadu National Park and the Fortescue River region, seems related to the abundance and distribution of resources, especially water, which determines whether the separation of neighbouring packs is due to temporal or spatial factors.

In the Fortescue and Kakadu wilderness regions, most territories apparently have separate and adequate water sources and hunting grounds, so that there is minimal competition between neighbouring packs. This spatial separation explains why there is relatively little contact between neighbouring packs, including howling, and why territory boundaries are stable over long periods.

By comparison, although food resources allow the survival of many dingoes in the Simpson Desert, water resources are rare and have to be shared. Howling and other aspects of communication between packs are therefore more pronounced so that the temporal separation of rival groups is more or less harmonious.

Dingo groups in these 'wilderness' areas tend to be large and remain together all year (which denotes greater stability), in contrast to dingo society in disturbed pastoral areas. There, the smaller social units and greater variation in group size throughout the year is mostly due to dingo control and other human disturbance.

The social organisation of most other canids, particularly coyotes, is similarly highly variable and generally determined by aspects of food availability and level of persecution by humans. Pack size in most other canids is also similar to dingoes, except African hunting dogs and dholes where packs up to 100 and 28 members, respectively, have been recorded. For dholes, two or more packs may assemble to form clans of 40–100 members.

THE SOCIAL GROUPINGS OF WILD DINGOES IN ASIA

Few data for these dingoes are available. Two groups of dingoes, each comprising 4–5 animals of both sexes, were observed over 6 weeks within and near Songkhla, a seaside town in southern Thailand. Animals of each group were exclusively associated with particular 'open air' restaurants where they were never fed but scavenged food scraps thrown aside by the restaurateurs and their patrons. Nobody, at least not the restaurateurs, claimed ownership of the dingoes. The dingoes of each group moved around the restaurants and other areas, more or less together, and their movement pattern was also fairly regular. They usually camped on the beach (see Plate 4).

On most evenings, soon after dark, both packs came together at what appeared to be the territorial boundary. The animals seemed very tense, and held dominant aloof poses with much whining, but there was seldom contact between pack members. They milled around for several minutes before departing.

However, there were frequent altercations between each pack's members and there most definitely was a 'bite order' (dominance hierarchy) enforced by a code of behaviour similar to that of Curly's Mob, although the scars and wounds suggested that this was not always adhered to. The dominant animals in each group had self-confident postures, often carried their tails high or curled over their backs, and always took precedence over others at food sources, or easily stole food from subordinant animals, who revealed their lowly positions by cringing, flattening their ears and licking the mouths of the dominant dingoes.

Similar interactions between dingoes have been recorded in other regions of Asia, from Myanmar to Sulawesi. Such aspects of the social organisation of dingoes in Asia resemble those of dingoes in Australia, especially packs living in relatively pristine areas such as the Simpson Desert. This, of course, is not surprising, considering the evolutionary relationships and connections between dingoes in Asia and Australia described in Chapter 1.

THE EVOLUTION OF DINGO SOCIETY IN AUSTRALIA

Like many other animals and plants in Australia, the life history of dingoes is shaped by drought. In particular, survival strategies are adapted to cope with the unpredictable onset, intensity, duration, and frequency of Australia's inevitable droughts. In fact, there may only be one predictable aspect of droughts: they will occur.

For dingoes in pre-European times, droughts meant huge changes in food and water sources, compared to the situation today where carcasses and artesian waters buffer dingoes through droughts (see Chapter 9). In those early times, food and water supplies alternated between periods of rarity and abundance, and the social behaviour of dingoes had to change from that of their wolf forebears to find and make the most of essential resources.

Such behaviour included the defence of their hunting grounds and scarce water sources. Even though the major prey of dingoes may have been small and easily captured by one dingo hunting alone, groups of closely related individuals would have been most efficient at gaining access to drinking water and at securing large prey (e.g. red kangaroos). Such prey often died near water during drought and usually took several days to eat. Waterholes are focal hunting points because animals have to drink there, especially large prey, and small species survive better in the moist habitats near water.

It is therefore likely that the primary function of dingo packs today is to defend hunting areas and other essential resources (water, carcasses, shelter etc); but as a spinoff, they can also kill large prey, particularly in areas such as the

Fortescue River region where packs are maintained to exploit the large macropodid prey that predominate there.

During extended drought in pre-European times, when all prey became scarce, dingoes could not maintain packs and members hunted separately, but probably maintained their former associations in the form of loose-knit tribes that shared a common area (the formerly defended territory). This notion is supported by the way pups display egocentric behaviour very early in life (about 4 weeks) that develops fully when they become independent.

Human activities such as pastoralism and associated dingo control (Chapter 8) are therefore probably not responsible for fracturing dingo packs, since this already occurred in response to droughts, as described. However, they may have extended the duration of this tribe structure. Also, the recent increase in artesian waters and drought food (cattle carcasses) has probably decreased the necessity for dingoes to maintain packs and defend water sources, even in droughts.

This hypothesis explains why dingoes still run in stable packs in the Simpson Desert today, even though their major food supply is small (rodents and rabbits). The packs maintain access to water, but still hunt independently for rodents or large prey such as kangaroos and feral stock, and still defend the carcasses of large kills. Similarly, pack formation of coyotes during difficult conditions in snowy winters appears to be an adaptation for the defence of carrion food, rather than the acquisition of live prey. This flexible behaviour resembles that of lion prides in Africa; members of prides hunt alone or cooperatively, even for the same prey species. Studies have shown that hunting success per individual lion does not increase with group size, so why do lions maintain prides? The reason is that their hunting grounds are quite small and local, usually around waterholes and river crossings, and it takes a pride to defend these areas (and kills) not only from other lions but from other predators such as hyenas and African hunting dogs.

CHAPTER 7

FEEDING ECOLOGY

This chapter examines what dingoes eat in different climatic regions of Australia and Asia and how they catch prey.

DINGO DIET IN AUSTRALIA

In the 20 years between 1966 and 1986, eight major federal and state studies on dingo diet were conducted in six climatic regions of Australia — from the northern wet-dry tropics, to the central hot arid deserts, to the cool south-eastern mountains. Collectively, these studies by federal and state government bodies provided 12,802 stomach and faecal samples to analyse the variety and relative amounts of prey eaten by dingoes.

For the combined six regions (essentially representing Australia), 177 prey species have been identified, and are listed in Appendix C. Almost 75% of prey eaten are mammals (72.3%, 71 species; see Figure 7.1a). The remaining quarter comprise birds (18.8%, 53 species), vegetation (3.3%, mainly seeds), reptiles (1.8%, 23 species) and an assortment of insects, fish, crabs and frogs (3.8%, 28 species).

The relative proportions of mammals, birds, reptiles and other prey that dingoes eat are remarkably similar throughout Australia (Figure 7.1b). Only in the coastal regions (north and south-east) are more birds eaten, and more reptiles are eaten in central Australia.

Almost 80% of mammals may be categorised as medium-sized or smaller (Figure 7.2a); whereas only 20.3% are large (see Box 7.1 for definition of categories). However, the relative proportions of the various sizes of mammal prey varies considerably across the six regions (Figure 7.2b), although medium-sized mammals usually predominate.

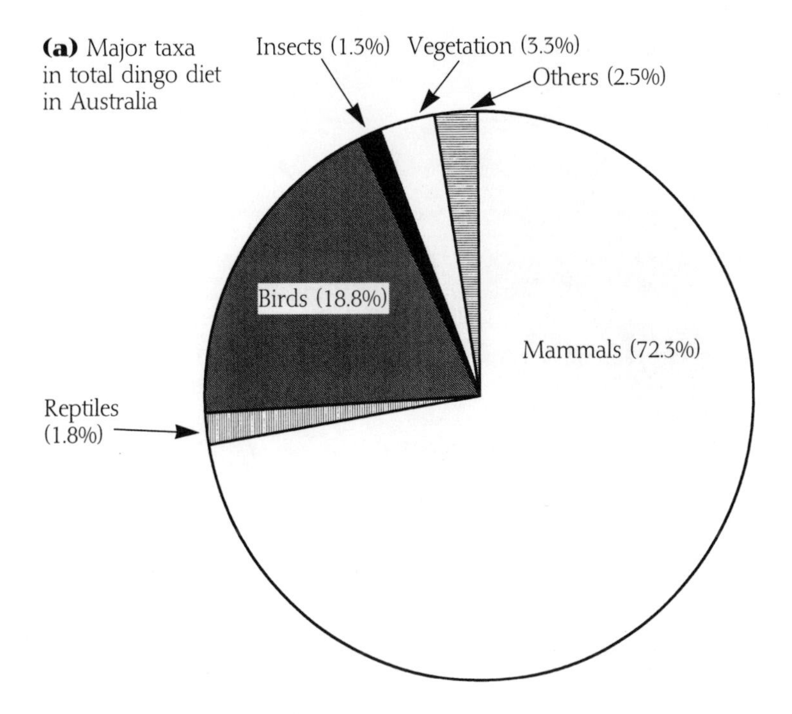

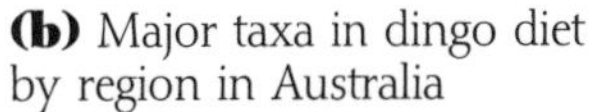

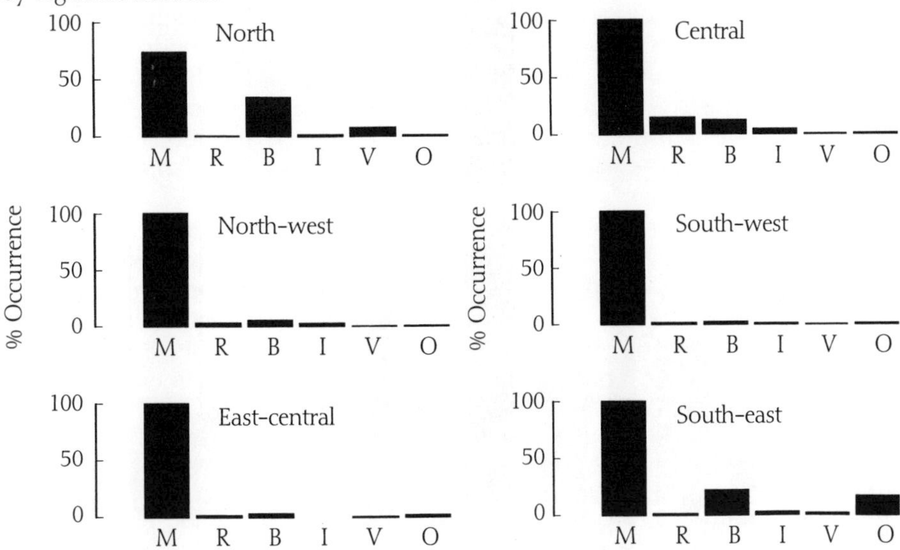

Figure 7.1
The relative amounts (% occurrence) of the major taxa in dingo diet indicated by 12,802 samples collected 1966–86. (a) Overall, almost 75% were mammals comprising 71 species; and (b) this pattern was true for 6 of Australia's environmental regions. M=mammals, R=reptiles, B=birds, I=insects, V=vegetation, O=all other dietary items.

(a) Size of mammalian prey in total dingo diet in Australia

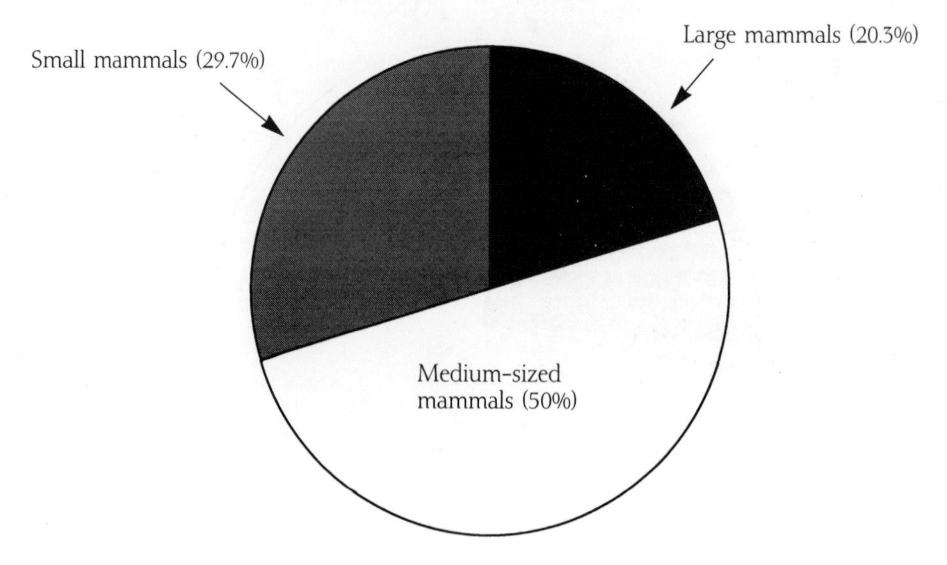

(b) Mammalian prey size by region in Australia (key as above)

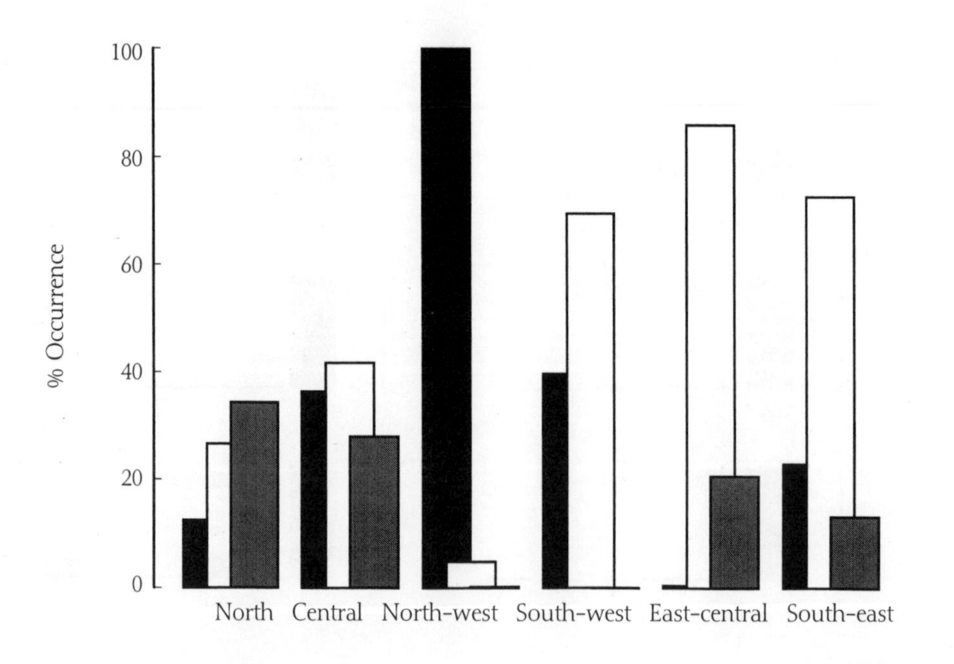

Figure 7.2
The relative amounts (% occurrence) of small, medium-sized and large mammals eaten by dingoes (a) throughout Australia, (b) by region. Medium-sized mammals, mainly wallabies, predominate in the dingoes' diet in most regions.

▲ **Plate 1** Despite their unique association with dingoes throughout the world over the past 10,000 years or so, humans rarely included them in their oral and written records. However, the Australian Aborigines regularly included dingoes in their stories and painted them in rock shelters; as shown by this rock painting from northern Australia.

▲ **Plate 2** Attaching a unique colour-coded collar to a dingo in central Australia. This will enable researchers to identify this animal later, and provide data on social behaviour and movements. Other captured dingoes were 'drugged' before handling and fitted with radio transmitters on collars. The people in this photograph, Peter Hanisch (left) and Harry Wakefield, were the stalwart technicians in CSIRO's dingo study in central Australia in 1966–75.

▼ **Plate 3** This male dingo, from northern Sulawesi, shows the ginger coat colour most common in dingoes throughout Asia (38–85%) and Australia (74%). Its short coat and relatively wide ears are typical of dingoes living in hot tropical habitats.

▲ **Plate 4** Part of a pack of 5 members, these dingoes are hunting small burrowing crabs along a beach in southern Thailand. This red-ginger variant is fairly common in Thailand (26% of 'ginger' dingoes) but is less common in Australia.

▶ **Plate 5** A black-and-tan dingo from central Australia. The white points and tan cheek spots on this female are typical. This colour variant forms 12% of the total Australian population but is not evenly distributed throughout the country. Another colour variant of pure dingo, solid black, is rare in Australia but relatively common in Asia.

◀ **Plate 6** The white dingo, a variant that forms 2% of the total Australian population, is not albino. As for most dingoes sampled for scientific research, this female from central Australia was foot-trapped then quickly and humanely killed by shooting in the neck so that blood could be collected and the skull was not damaged.

◀ **Plate 7** A hybrid showing the sable pelt coloration (6% of the total Australian population). This female from central Australia shows extensive black coloration along the dorsum and sides; many people mistakenly believe that this variant results from the mating between a dingo and an Alsatian. Another variant of sable shows only the dark dorsal strip.

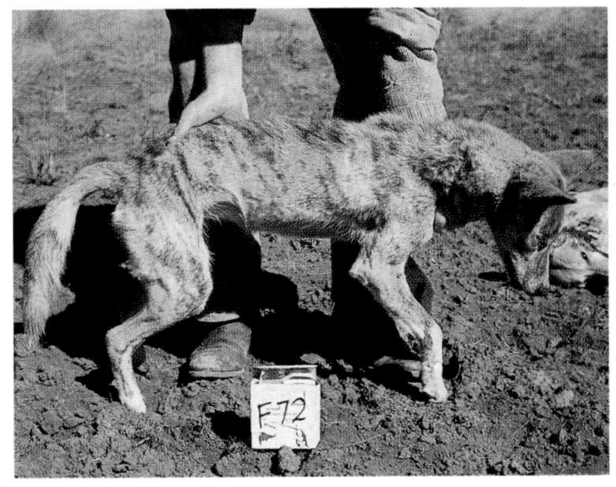

▲ **Plate 8** Brindle hybrids form 3% of the total Australian population. This female is from central Australia. In Victoria, sightings of similarly coloured animals sometimes give rise to the forlorn hope that thylacines still exist in the bush.

◀ **Plate 9** A hybrid male from a mating in captivity between a ginger dingo male and a blue heeler female. This animal closely resembles a pure dingo and shows how hard it is to identify the hybrids of dingoes and dingo-like domestic dogs from external characteristics such as coat colour and body shape. Only skull measurements can reliably identify such hybrids, although the technique of DNA fingerprinting may prove useful in the future.

▶ **Plate 10** Patchy hybrids are uncommon in Australia (3% of the total population) compared with Asia (up to 37%). This male from central Australia shows a large white 'collar' that extends from the neck to the shoulders, face, chest and front legs; there is also a small white patch on the loins and the white tail tip is larger than it is in pure dingoes. The pattern of patches in hybrids varies greatly though the most common variants are ginger-and-white and black-and-white.

▲ Plate 11 In Asia, dingoes and humans usually have a commensal relationship: both derive benefit from the close association, for example, food or protection, but neither depends on the other for survival. This free-ranging dingo in northern Thailand is scavenging food scraps from a market stall. Other dingoes are regularly fed by householders in return for warning them of human intruders.

▲ Plate 12 Dingoes, including hybrids and dogs, are regularly eaten throughout Asia and the Pacific islands. In some regions dingoes provide the major protein component of human diet and is preferred to beef, pork, chicken and fish; in other regions, they are eaten for medicinal purposes, such as curing fevers and stomach ailments. At this abattoir in north-eastern Thailand, dingoes are being butchered into various cuts prior to transport to a morning market. Coincidentally, these dingo abattoirs provided a reliable source of skulls and other samples for comparison with Australian dingoes.

▲ **Plate 13** Six pups, about 5 weeks old, at a typical natal den in central Australia — an enlarged rabbit hole in a warren still occupied by rabbits. In another month or so these pups will play and rest in the nearby dry creek bed and adult dingo relatives (helpers) will coach them in stalking the rabbits that apparently have been reserved for that purpose. At about 3 months the pups will be based at separate rendezvous sites in the pack territory and will finally be weaned at a watering point.

▲ **Plate 14** Dingo society is based on aggression, and each pack has a distinct dominance hierarchy, particularly for males. However, in stable packs many disputes are settled by ritual behaviours instead of outright fighting. In this photograph the upright dingo is asserting its dominance over the 'passively submissive' dingo, which is characteristically recumbent with its tail tightly tucked between its legs, ears flattened and tongue protruding.

▲ **Plate 15** A dingo pack at a shared waterhole in the Simpson Desert. Although these dingoes appear relaxed while drinking, they are constantly on their guard against rival packs. This is indicated by their alert ears and upright tails in dominant positions.

▲ **Plate 16** Dingoes howl to attract mates and repel rivals. Chorus howling by dingo packs indicates group size, and is achieved by members howling together and using different variations of the three basic types of howls: plateau, inflexion and yelp (Chapter 5). For large groups, some members may start with the same howl but shift to a different type of howl (varying the pitch and duration); the more dingoes howling, the more types of howls are used and this is how rivals 'count' the numbers. Chorus howls probably also help to identify packs and are mostly used to repel rivals from essential resources. In this photograph, the East Pack from Alka Seltza in the Simpson Desert, are responding to their rivals, the West Pack (Chapter 6), who were howling from a nearby watering point that both packs shared.

▲ **Plate 17** In arid Australia, watering points and associated carcasses are usually shared between adjacent (territorial) dingo packs so that dingoes are wary when they approach them. Once certain that the coast is clear, packs perform a ritualised display of cautionary bravado by storming around the water in a compact group, hackles raised and tails curled over their backs, dust rising. In this photograph, the East Pack from Alka Seltza in the Simpson Desert checks out the carcass of a bull that was shared by about 23 dingoes.

▲ **Plate 18** In stable packs successful mating usually occurs only between the dominant (alpha) pair of dingoes. Although other mature females come into oestrus and are mated by the alpha and sometimes by other males, their pups are usually killed by the alpha female soon after they are born. In this photograph, 'Whiteface', the alpha male of the East Pack at Alka Seltza in the Simpson Desert is assessing the reproductive status of a subordinate female. Another subordinate male looks on. A few days later this female was mated by 'Whiteface' but her pups were never seen; only those of the alpha female's were in evidence.

▲ **Plate 19** Dingoes, particularly males, ritually urinate and defaecate on scent-posts as part of their communication. Scent-post urination follows a seasonal trend that is linked to the breeding season and may indicate the reproductive status of potential breeding partners. At other times of the year scent-posts are sometimes used to indicate which shared hunting grounds and/or watering points are being used. Here, a male is performing a 'raised leg urination' on a small bush. The urine (and its associated message) are deposited on an elevated object so that the smell carries further. Because of its rarity, this conspicuous scent-post probably also provides a visual guide for other dingoes.

◀ **Plate 20** In the Barkly Tableland, the dingo mating season coincides with the cattle calving peak and this coincidence often contributes to the deaths of many calves. Calves are attacked by sexually excited dingoes and killed because they cannot behave like submissive dingoes and thus appease or divert the aggression, as normally occurs in dingo society. As indicated by the skinned calf shown here, calves killed this way are rarely eaten but even if this one had lived and reached an abattoir, the meat would have been downgraded due to the scars from dingo bites.

▲ **Plate 21** Most dingo populations are 'controlled' by human activities, or are naturally regulated by social behaviours such as infanticide. Occasionally, however, pathogens and diseases devastate local populations — distemper in arid Australia, for example, or heartworm in the coastal wetlands of northern Australia. This female dingo was healthy and her radio-collar fitted snugly when she was captured only one month before this photograph was taken. Although easily captured food (dusky rats) was abundant, the presence of over 100 adult worms in her heart apparently restricted her activities so severely that she starved to death.

Diet by region in Australia

In the tropical coastal regions of the Northern Territory, dusky rats, magpie geese and agile wallabies together form over 81% of the diet (Appendix C) and there is a regular pattern of predation on them. Most geese are eaten during the dry months (May-October) of each year when newly fledged young are available. Dusky rats also predominate during the dry months, particularly during years of high abundance, about every third or fourth year. In the meantime and during the wet months, agile wallabies form the bulk of the diet.

In central Australia, the introduced rabbit has, paradoxically, become an adequate substitute for native game, particularly during runs of high rainfall years (see Chapter 9). During droughts, dead cattle provide most of the diet. Further north, on the Barkly Tableland, where rabbits do not occur, no single native species predominates in the diet except for long-haired rats when they form huge plagues (about once every 9 years) and overrun the countryside.

The Fortescue River region in north-west Australia is the only region where large native mammals, red kangaroos and euros, predominate in the diet, probably because of the

Box 7.1
Mammalian prey categorised by body size and ease of capture

Studies of dingo diet in different regions of Australia indicate large differences in the species composition of prey eaten and in the hunting tactics used to catch them. This is because of the great variation in the types of prey available, the relative abundance of the various species of prey, and the nature and stability of the environment. However, the body size and anti-predator behaviour of prey can be categorised and correlated with dingo hunting tactics to allow useful comparisons of dingo diet.

Small and medium-sized mammals are categorised as adults that weigh up to as much as an adult dingo weighs (about 15 kg), and can be killed by one dingo hunting alone. Small mammals weigh up to about 0.5 kg and medium-sized mammals weigh between 0.5 kg and 15 kg. A large mammal, on the other hand, is categorised as one that weighs considerably more than a dingo does and requires a pack of dingoes hunting co-operatively to kill it. The young of large prey often form the bulk of this category, and, although some may weigh considerably less than a dingo, they are included because dingoes usually have to overcome the formidable defences of adult parents before they can kill their young.

paucity of small and medium-sized mammals, particularly rabbits. Further south, on the Nullarbor Plain, red kangaroos and rabbits are abundant and dingoes eat twice as many rabbits as kangaroos.

In the cool temperate mountains of eastern Australia, the medium-sized wallabies (swamp wallaby, red-necked wallaby) predominate in the dingo's diet in the lower slopes, and the wombat predominates at higher altitudes. Possums (brushtail, ringtail) are also commonly eaten when they come to ground to feed.

ARE DINGOES GENERALISTS OR SPECIALISTS?

Although a large range of prey is eaten throughout Australia — at least 177 species — almost 80% of the dingo's diet comprises only 10 species. In order of greatest frequency these are: red kangaroo, rabbit, swamp wallaby, cattle, dusky rat, magpie goose, brushtail possum, long-haired rat, agile wallaby and wombat. Of these, only cattle (mostly as carrion) are eaten in each of the six regions studied. Given the huge range of potential prey, the narrow range of species that dingoes focus on strongly suggests that they are specialists, not the opportunistic generalists they are often assumed to be. However, in terms of hunting strategies, the generalist tag applies because of the broad range of hunting tactics that solitary and co-operating pack members use (see page 113).

That dingoes seem to ignore many potential prey species is surprising and difficult to explain. For example, why are so few reptiles eaten? At Kapalga, in Kakadu National Park in northern Australia, 70 species of reptiles have been recorded. Many species, such as frilled lizards and goannas, are very common, accessible and don't seem to display formidable anti-predator behaviours; yet only 8 of these individuals were recorded in 6,722 faeces over 7 years of study! In central Australia, lizard diversity and abundance is even greater, but apart from the regular consumption of one species (the central netted dragon, 7.8% occurrence), lizards are largely ignored even during drought when dingoes may be starving.

Search image

One of the reasons why apparently obvious prey are 'ignored' may have something to do with the 'search image' that dingoes have of prey, particularly in regions where prey diversity is usually low. On the Barkly Tableland of the Northern Territory there are sporadic plagues of the long-haired rat, about one in every 9 years. Once up and going the rats overrun huge tracts of countryside for several years before disappearing to largely unknown refuges. In the early 1970s huge rat plagues engulfed the entire Barkly Tableland (and beyond) and several generations of dingoes survived on an almost exclusive diet of long-haired rats.

When the rat populations suddenly crashed due to drought, some dingoes scavenged cattle carcasses, but most dingoes died of starvation or succumbed to canine hepatitis. Even when flush times returned, the surviving dingoes apparently did not take full advantage of potential prey species such as songlarks, pipits and other ground nesting birds which were then breeding in great profusion. It seemed that dingoes had such a specific search image for long-haired rats that they were unable to switch to other prey.

In the Fortescue River region of north-west Australia, the dingoes that moved from the hills to the sheep paddocks seemed to take a long time to recognise sheep as potential food; they continued to hunt their 'traditional' prey — red kangaroos — despite the presence of many sheep. Eventually when these dingoes did attack sheep they were inept hunters, or sometimes they killed sheep before going off to kill kangaroos. Conversely, dingoes that had been raised in the sheep paddocks preyed almost solely on sheep and were inept in killing kangaroos.

Similarly, in other species, familiarisation means that one type of prey is preferred over others that have not yet been encountered by the predator. In experiments with captive snakes and birds, the primary food was preferred to others offered subsequently, so that switching did not occur and the animals starved to death.

This search image notion may apply to dingoes only in regions where potential prey diversity is fairly low and constant over several consecutive years, because in other regions of high prey diversity dingoes are justifiably renowned for their adaptability. For example, dingoes living in the gibber country on the western edge of the Simpson Desert normally subsist on a variety of small vertebrates. During one drought year when these prey were rare, dingoes capitalised on large flightless grasshoppers which, in character with that region, suddenly appeared, and perhaps have rarely been seen since. Even more remarkable was how the dingo mothers there regularly but independently moved their litters of young pups to temporary den sites to coincide with the areas of greatest grasshopper activity, as described in Chapter 6.

ARE DINGOES OPTIMAL FORAGERS OR EFFICIENT SATISFICERS?

Since dingoes generally forage for patchily distributed food, rather than randomly encountering prey in a patch, the prey model (Box 7.2) is probably the most appropriate one for correlating dingo foraging with the key predictions of optimal foraging theory.

Box 7.2
Foraging theory

It is assumed that a predator allocates its resources of time and energy in the most profitable or 'optimal' manner. Optimisation models propose that a predator maximises net energy gain per unit hunting time per unit energy requirement, and models can be *dynamic* or *static*. Dynamic decision models try to formulate a sequence of decisions that allows a predator to alter its decisions from time to time; decision variables include experience, hunger, body size and previous decisions. Dynamic models usually focus on tradeoffs — for example, conflict between feeding and territorial defence.

Static models involve simple, non-sequential decisions, and include the classical (conventional) models most frequently used in foraging theory. Classical models aim to maximise the average rate of energy intake E_f/T_f where E_f is the net energy gained from foraging in time T_f, the sum of search and handling times. There are two classical models: prey (diet), and patch. Both assume a sequence of search — encounter — decide. The fundamental difference between these models lies in the decisions they analyse, not the nature of prey encountered. Prey models analyse whether a predator should eat a particular prey type or reject it and continue searching. Patch models analyse how long a predator should stay hunting in a particular patch.

A particular prey provides a fixed amount of energy (e), and there is a fixed amount of time to pursue, capture and consume that prey, together comprising handling time (h). For prey models, the profitability of a particular prey type is determined by the value of e/h. The principal results and predictions of the prey model are:

1 Prey types are always taken upon encounter (p = 1) or never taken upon encounter (p = 0), where p is the probability that a prey type will be attacked upon encounter. This is known as the 'zero-one rule'. One prediction is that when all prey are sufficiently abundant, predators select only the most profitable prey type(s). Another prediction is that prey types are completely included in or completely excluded from the optimal diet so that there can be no partial preferences whereby particular prey are sometimes included in the diet.

2 Predators rank prey types by their profitability (e/h ratios). Prey types are added to the optimal diet in order

continues ...

of their ranks so that $e_1/h_1 > e_2/h_2 > e_3/h_3 > ...... e_n/h_n$. One prediction is that predators will not switch between two prey types to consume a disproportionate amount of the most common of them.

3 The inclusion of a prey type in the optimal diet is independent of its own encounter rate (abundance), and depends only on the absolute abundance of more profitable prey. Thus, as prey abundance declines, the diversity of prey in the diet should increase with prey added in order of profitability (i.e. predators will generalise). Conversely, an increase in overall prey abundance will lead to greater specialisation on prey of high profit.

The principal result of the patch model is the 'marginal-value condition' whereby the patch residence time is set so that the instantaneous rate of energy gain at leaving it equals the long-term average rate of gain in that habitat.

Prey model

At a study site in arid central Australia (site 5, Table 2.1), there was enormous variation in the abundance of most prey between runs of flush and drought years, but dingoes mostly ate rabbits (ranked first for 7 years, mean 57% occurrence) irrespective of their abundance and climatic conditions; this preference supports predictions 1 and 2 of the optimal foraging prey model. Rabbits are the most profitable prey because they can be captured by one dingo hunting alone and one rabbit (mean body weight, 1 kg) also satisfies the daily energy requirements of an adult dingo.

During drought, when the abundance of most prey declined, 10 new species were added to the range of dietary items (supporting prediction 3), and were dropped after the drought broke (supporting prediction 1). Except for rabbits, most of the 26 prey species eaten were deleted and/or added to the diet at least once during the 7-year study; such variable consumption (partial preferences) contravenes prediction 1. However, because dingoes often had a range of profitable prey to choose from (in flush seasons), and since the threshold encounter rates of prey seemed to vary, such partial preferences can be expected.

In the northern tropics at Kapalga (site 1, Table 2.1), the abundance of many prey fluctuated with the wet and dry seasons. The prey staples there — rats, geese and wallabies — together made up 65–95% of dingo diet annually over the 7-year study; such consistent preference supports predictions 1 and 2. Magpie geese were probably the most profitable prey in terms of benefit gained per effort because one fledgling or adult goose (weighing up to 3 kg) is more than sufficient to satisfy a dingo's daily energy requirements. In response to an experiment that increased the abundance of most prey, the consumption of wallabies increased concurrently with a decrease in the overall diversity of prey in dingo diet; this specialisation on prey of high value satisfies prediction 3.

Patch model

Some data suggest that dingoes probably also conform with the predictions of optimal foraging patch models. Figure 9.2 shows, for example, an inverse relationship between floodplain fauna (rats and geese during the dry months) and forest fauna (wallabies and possums during the wet months). This inversion occurs because dingoes are obliged to spend more time hunting in the margin and forest habitats when the plains are flooded, and consequently consume more wallabies, possums and other prey. However, as soon as the plains begin to dry out, and before the next annual flooding occurs, dingoes hunt mainly on the plains for rats and young geese, which are more accessible then.

Thus there is essentially a 2-prey system (geese and rats) on the floodplains in the dry season. The data in Figure 7.3 indicate that dingoes did not switch between geese and rats and did not contravene prediction 2. Instead, they tended to eat more rats when they were scarce, and radio-tracking indicated that they also focused on local pockets of relatively high rat density, which again supports the notion that dingoes were patch hunting.

During flush years in central Australia the relative abundance of rabbits and rodents (both profitable prey) sometimes fluctuated independently and dingoes apparently consumed disproportionate amounts of the most common of them. That appears to contravene prediction 2, but there are insufficient data to analyse switching for this region.

Tradeoffs between foraging and other activities

Overall, data on the dingo's foraging behaviour in both habitats mostly concurred with the key predictions of optimal foraging theory. The acceptance of these predictions, however, was not unequivocal, partly because of the limitations of using stomach contents and faeces to test predictions; it is impossible to determine if changes in diet reflect changes in prey choice or patch choice. Also, the Australian climate causes annual or intermittent fluctuations in the abundance of many prey. Dingoes are very mobile and regularly encounter several kinds of prey simultaneously so that many

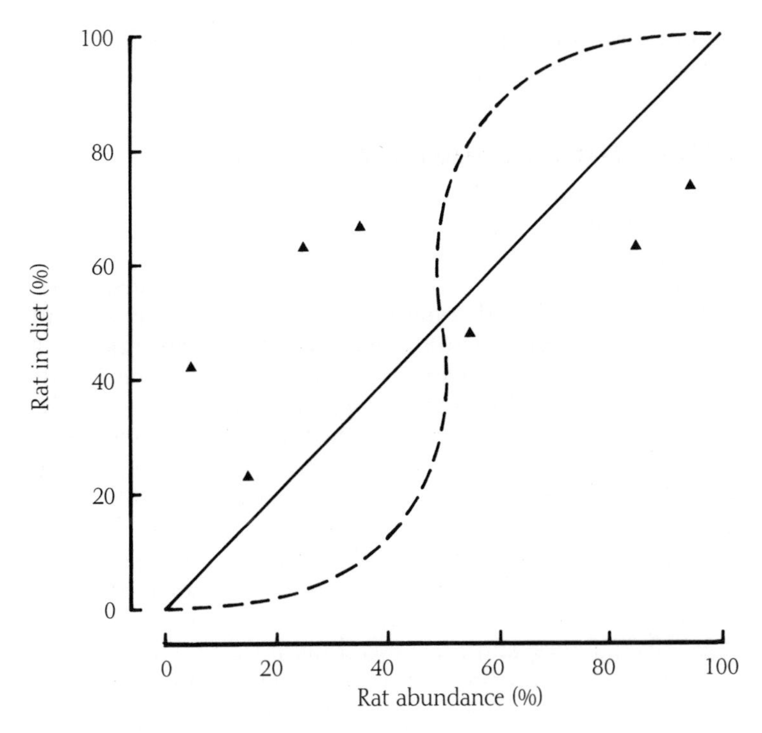

Figure 7.3

Do dingoes switch between prey types? At Kapalga, Kakadu National Park, they mostly hunt on the floodplains in the dry season where they have what is essentially a 2-prey system: dusky rats and magpie geese, both profitable. The triangles show amounts of dusky rats in dingoes' diet as a function of their abundance over 7 years. These diet and abundance data were measured as the percentages of the highest values recorded for % occurrence in faeces and numbers of rats trapped per 100 trap-nights, respectively. The solid line indicates the theoretical function expected if preference is consistent and the broken line indicates the expected theoretical function if dingoes switched preference to whichever prey is most common. Dingoes did not switch, nor display a consistent preference but tended to eat more rats when they were rare.

of the model's assumptions are not always met. These assumptions require sequential encounters and hunting in uniform patches with fixed hunting strategies.

In the dingo's natural world, there must be tradeoffs between foraging and other competing activities such as territorial defence, competition, wariness, dominance, satiation, defence of resources (breeding mates, water, carcasses) and avoidance of physical stress. Dingoes cannot always adopt the strategy that would maximise their foraging profits even though natural selection may demand it. Dingoes adopt whatever feeding strategy is most efficient at the time, to gain at least a threshold level of nutrient. Accordingly, it can be said that dingoes are efficient satisficers where their foraging usually involves a large, fluctuating set of dynamic models.

DINGO DIET IN SOUTH-EAST ASIA

In Thailand plant carbohydrate comprised 53.2% of 216 observations of feeding dingoes. The main items were cooked rice, raw banana, raw coconut, cakes and unidentified types of vegetables, fruit and confectionery (Table 7.1). Animal protein was a relatively minor component (8.3%) and included raw or cooked fish, chicken, cow, crab, prawn and dog. Many meals (38.4%) were of mixed animal and plant items, mostly cooked rice with small pieces of chicken, fish or unidentified meat; these usually came from take-away food outlets or households (see Plate 11).

Very few dingoes in Asia live totally independent of human influence. In rural areas of Thailand and in north Sulawesi, for example, 52 observations were made of dingoes hunting insects, rats, lizards and other live prey along roadsides, rice paddies, and in forests; but all those dingoes were also associated with and probably obtained most of their food from households in nearby villages.

Table 7.1

The percentages of animal protein and plant carbohydrate in dingo diet in Thailand and Australia.

The relatively greater intake of carbohydrate in Asia explains the smaller body size of dingoes there.

	Animal protein %	Plant carbohydrate %	Mixture (mostly carbohydrate) %	No. of samples
Thailand				
North	7	53	40	68
North-east	7	68	25	60
Bangkok	10	43	47	88
Mean Thailand	**8**	**53**	**39**	**216**
Australia				
North	98	2	0	6722
Central	99	1	0	285
West	95	5	0	146
East	99	1	0	1993
South-east	98	2	0	2171
Mean Australia	**98**	**2**	**0**	**11317**

These data suggest that for dingoes in Asia the main diet item is plant carbohydrate, supplied directly or indirectly by the people with whom they have a commensal relationship. In contrast, dingoes in Australia are wild and hunt live animal prey so that their major diet item is animal protein (>95%; see Table 7.1). These dietary differences may well explain the smaller body size of Thai dingoes and support the suggestion that Thai dingoes warrant classification as a distinct subspecies.

DINGO HUNTING TACTICS AND SUCCESS, AND THE ANTI-PREDATOR BEHAVIOUR OF PREY

In general terms, prey are targeted and assessed in terms of their ability to defend themselves and their ability to injure the dingoes severely. Then the dingoes attack them with an appropriate tactic. Tactics used to catch prey may differ from those used to kill them. These differences and the ultimate degree of hunting success are related to many factors, including the number of attacking dingoes, the age and size of prey, the anti-predator behaviour of prey, hunting terrain, dingo social status, prey and dingo health, and individual dingoes' hunting experience. Some of the dingoes' hunting tactics and their success are described as follows.

Hunting large kangaroos: tactics

Throughout Australia various species of large kangaroos (mean adult weight 17–66 kg) are the most commonly killed large prey and the tactics dingoes use to catch them are generally the same: they sight them, bail them up, and then kill them.

Packs of dingoes are more than three times as successful at bailing up kangaroos and more than twice as successful at killing them than dingoes hunting by themselves are. For example, in the Fortescue River region in north-west Australia the success rate for single dingoes bailing up and killing red kangaroos and euros was 5.5% and 33.3% respectively. The equivalent success rate for groups of three or more dingoes hunting together was 18.9% and 74.2% respectively.

The advantage of numbers when *pursuing* large kangaroos is that the leading dingo often makes the kangaroo change direction into the path of other dingoes involved in the chase, who in turn are skilled at cutting corners; the combined effort soon exhausts the quarry and it is easily bailed. This technique is similar to that used by wolves, hyenas and African hunting dogs. Another variation that African wild dogs use always and dingoes sometimes is the 'relay hunt' in which exhausted leaders are replaced by following pack members.

The advantage of numbers when *killing* large kangaroos is that they can maintain relentless pressure on the quarry after a successful chase. Since the leading dingoes become exhausted during the chase, the killing is done by other members of the group coming along behind them. The quarry can also be attacked simultaneously from several directions.

Autopsies of kangaroos killed by dingoes suggests two patterns of attack: (1) nipping or hamstringing the hind leg to slow the kangaroo sufficiently to attack its throat (on adult and juvenile kangaroos); (2) running alongside and biting the dorsal thorax and neck regions (of juveniles and small adult females).

Hunting success in open and closed habitats: anti-predator behaviour

Dingoes are probably more successful at hunting large kangaroos in fairly open areas, particularly arid areas where kangaroos tend to concentrate around permanent sources of water, especially during droughts. This can lead to surplus killing. For example, at a dam enclosed by a wire fence in central Australia, dingoes apparently learned to rush at drinking kangaroos, which made them panic and crash into the fence. Those that became temporarily immobilised were easily mauled or killed, and many more were killed than eaten.

In contrast, hunting success is less frequent where kangaroos can use the natural features of the terrain to escape pursuing dingoes or hamper their attack. In the dissected ironstone ranges of the Fortescue River region in north-west Australia, 62.5% of unsuccessful attacks by groups of dingoes on adult kangaroos were unsuccessful because the kangaroos backed up against natural barriers, preventing attack from the rear and reducing the attack to only one dingo at a time. Presumably also, the success rate of hunting grey kangaroos in the densely forested regions of eastern Australia is also low because of similar obstacles; but there are no data available.

The apparent differences between hunting success in open and closed habitats are reflected in the alert and flight distances that kangaroos use in these habitats. Eastern grey kangaroos alert to dingoes sooner than red kangaroos (means: 121 m and 150 m respectively) but both flee at similar distances (means: 105 m and 98 m respectively). On the other hand, being forced into a dead-end ravine can also prove a kangaroo's downfall. For example, the remains of many euros at Oodnappina rockhole in the western Macdonnell Ranges in central Australia suggest that individuals had been regularly trapped and killed by dingoes.

There are many records of adult kangaroos and wallabies taking to water to escape attack by dingoes. This ploy is successful if the quarry can swim across a river or lake but not always so if there is no escape route (as at a small dam or waterhole) because the dingoes often wait for the quarry to emerge and then continue the attack. Most records describe kangaroos standing chest-deep in water with one or more dingoes sitting on the bank watching, which suggests that the popular belief that kangaroos drown dingoes is unfounded. Most dingoes are probably too cautious and only the young, inexperienced dingo or even the drover's kangaroo hound are likely to pay such a penalty for entering the water.

Hunting medium-sized wallabies and other small macropodids

The tactics dingoes use to chase and kill small and medium-sized macropodids (mean adult weight 3–20 kg) are similar to those they use for large kangaroos, but some differences are apparent. Dingoes hunting alone rely more on scent than sight to trail the quarry, so they are often several hundred metres behind them and the chase may last for several hours. The success rate of lone dingoes hunting medium-sized macropodids is probably greater than it is for lone dingoes hunting large macropodids, but there are no data to confirm this.

In the moist forested mountains of north-eastern New South Wales, it was observed that female swamp wallabies pursued by dingoes eject their pouch young (>c. 800 g) apparently because they are a burden. However, there are no data on whether or not more females escape by this ploy; it is only known that the population age structure of wallabies is altered.

Ages of macropodid prey

Most studies indicate that the young of large and medium-sized macropodids are killed more frequently than adults. In arid north-western New South Wales, almost all red kangaroos killed by a pack of five dingoes around a dam were juveniles less than 18 kg (96.4%) and most were females (81.3%). In the moist forested mountains of north-eastern New South Wales (Georges Creek Nature Reserve) where swamp wallabies are the major prey (44% of dingo diet), dingoes also indicated a strong preference for dependent juveniles (32%), but not for one sex or the other.

Several studies have remarked on how most large male red kangaroos are indifferent to the presence of dingoes, even when dingoes were attacking younger males and females amongst them. The studies concluded that large males are not routine prey. Sometimes, however, dingoes regularly and successfully tackle large males; for instance, they do in central Australia when drought reduces other main prey species. In one instance three dingoes were seen attacking an 'old man' kangaroo on the open plains. One dingo held a front position and the others darted in from the rear and inflicted nasty wounds on the hips and around the base of the tail. These dingoes were inadvertently driven off by human observers but they probably would have succeeded because the kangaroo was too exhausted to flee from the humans and soon died from its wounds.

In eastern Australia, a large male eastern grey kangaroo successfully defended himself from a dingo for over an hour by high-standing, kicking, tail-thumping and hopping towards the dingo. However, after the previous description, one wonders what might have happened if other dingoes had arrived on the scene.

Hunting large exotic mammals – cattle and buffalo

Nearly all attacks are aimed at young animals, from the newborn to subadults; healthy full-grown adults are rarely attacked. There are three basic attack tactics: (1) constantly harass a mother with a dependent calf; (2) actively spook a mob of cattle to separate the calves from the adults, then focus on the calves; (3) sit, wait and watch a mob of cattle or buffalo and investigate conspicuous or unusual behaviour by individuals. Dingo groups use all three tactics but dingoes hunting alone or in pairs mainly use the first tactic. The following examples illustrate these tactics as well as some of the anti-predator behaviour that cattle and buffalo use to defend their young.

The 'constantly harass' tactic

This tactic is simple: dingoes keep the prey constantly on the alert, so it eventually tires or relaxes its guard enough for a dingo to deliver a crippling bite.

Probably the most frequently recorded interaction of this tactic is that of a single dingo circling a mother cow or buffalo with its very young calf. In many cases the dingo appears to be just checking out the mother's 'maternal instincts' because she usually charges the dingo and drives it off. However, the dingo may persist or, more successfully, combine with one or more other pack members to attack the calf. When the calf is threatened, the mother is roused and occasionally charges the dingoes. This harassment continues for many hours, even days, with the dingo sub-groups alternating between harassing and resting. Whenever the dingoes elude the mother's guard, they bite the calf's hamstrings, lower back region, shoulders or ears. Eventually the calf is so maimed that it dies, or the mother becomes exhausted and has to leave the calf to drink or feed, and the dingoes can then kill it.

The maternal instinct of cows and their ability to defend their calves varies considerably; some mothers abandon their calves after the first attack, others defend them long after they have died, but many are successful. This is inferred from observations of many otherwise healthy calves with torn ears and/or scarred legs, and from the fact that up to 5% of meat at the Katherine meatworks in the 1970s was downgraded because of scarring from dingo bites.

The 'constantly harass' tactic is most successful for killing very young calves, but sometimes it succeeds with large animals. At Kapalga in Kakadu National Park a pack of six dingoes was observed chasing an 18-month lone male buffalo (estimated at 200 kg); the lead dingo was nipping at the buffalo's legs while the others followed, then another dingo replaced the leader. In this case, the end was not witnessed as the animals disappeared into the scrub, but the buffalo seemed very tired; its tongue was hanging out and it constantly bellowed. The remains of other buffaloes this size have indicated that they had been killed by dingoes.

Another example, from central Australia, involved a group of four dingoes which had targeted a small black calf in a herd of about 30 cattle. The dingoes operated in pairs and alternated between attempting to separate the calf from the herd and resting nearby. One pair managed to separate the calf and knock it over but its mother charged and drove them off, and the calf returned to the relative safety of the herd. As dusk approached, after about 130 minutes of observation, the herd stopped feeding and began to make their way to water. The calf could not keep pace, apparently because it had been injured by the earlier attack. Despite the mother's heroic attempts to protect the calf, the four dingoes harassed her long enough for one of them to deliver a killing bite.

The spook tactic

In arid Australia mobs of cattle drink from the widely spaced artesian bores. Here water is pumped into a dam or trough, often enclosed by a cattle yard. Dingoes take advantage of the congestion and confusion when cattle compete for the water to spook the mob, with the aim of separating a mother from her calf long enough for the dingoes to kill or disable the calf.

For example, during a moonlight observation from atop a windmill at Tieyon Station in northern South Australia, at 0230 h, about 50 cattle adults and calves entered a yard and began drinking at a trough. Suddenly five adult dingoes appeared and ran among the cattle. The cattle were immediately spooked and stampeded, bellowing loudly, around and out of the yard, with dust rising everywhere and probably adding to the confusion. In a matter of seconds only about 10 cattle remained in the yard; about 20 had run off to the east and 10 or more had gone north, with the dingoes running in amongst them.

Then the dingo group split. Two ran flat out to the front of the north-heading cattle, the other three cut in and out at its rear, but all became obscured by dust and trees at about 300 m from the bore. Meanwhile, the cattle in the yard had headed west except for one calf that bawled and ran about before trotting north after the cattle that were being attacked; perhaps their bellowing attracted it. Later that morning the fresh remains of two calves were found about 400 m north of the bore.

On another night at the same location, about 40 cattle were watering and five dingoes (perhaps the same ones) came running into the yard and spooked them. Several minutes later, after the dust had cleared, most of the cattle had run off except for five adult cows who were standing in a circle, facing outwards, to protect two very small calves in the middle. Two dingoes watching this group suddenly took off after the other cattle. This anti-predator behaviour that the cattle used to protect their young is also commonly used by buffalo in the northern coastal wetlands of Australia. This suggests that dingoes may also use the 'spook tactic' on buffalo but there are no records of their success at stampeding them.

The wait and watch tactic

This tactic involves observing the most vulnerable members of a herd, especially sick or weak calves, with minimal energy loss and minimal risk of injury. Dingoes (usually a group) simply sit, sometimes for hours, and watch the herd from a distance. They closely investigate only the cattle or buffalo herd members that exhibit unusual or otherwise conspicuous behaviour — anything that differs from what other members of the herd are doing at the time.

If such behaviour is due to an affliction or injury that makes an individual vulnerable to predation, the dingoes either attack directly, harass the quarry (tactic 1), or wait until it dies. If the individual turns out to be healthy, the dingoes promptly back off and resume waiting and watching for other potential quarry. Common examples of conspicuous behaviour are: mothers giving birth, individuals that are 'stationary' and not moving with the main body of the herd, calves running and frequently bawling, individuals that do not go in to drink with the herd, and individuals straggling behind the herd.

The following description is a typical example of how dingoes use the 'sit and wait' tactic. It involved a group of two adult dingoes and their five almost fully grown pups near the Finke River in central Australia. This group spent many hours each day, sitting in the shade and generally watching over the extensive river flats where about 200 cattle were grazing in several groups. Sometimes the dingoes drank at a nearby bore or fed from the carcass of an adult cow that had been killed by the station lessee (for his meat). There were many calves and more were born each day. Whenever calves seemed to straggle, or became separated and ran to catch up with their mother, the dingoes got up and investigated, but left as soon as it was obvious that the calf was healthy.

In one instance, a new-born calf was staggering behind its mother who had an extremely full udder. Although the calf attempted to suckle, it was apparent that it was receiving no milk, or very little, and was weak from hunger. When the dingoes first investigated, the mother charged and managed to ward them off. The dingoes resumed waiting and watching, but they regularly moved to the cattle mob to check this calf, and fed from the adult carcass in the meantime. After 4 days the calf died and the dingoes ate it.

Nursery groups

Both cattle and buffalo often leave young calves with adult nursemaids while the main mob travels up to several kilometres to feed or drink. A typical nursery group would consist of about 10 small calves, two bigger calves, and three adults. These nursery groups seem highly vulnerable to dingo attack, although they have only been seen during daylight hours which may partly explain why dingoes rarely attack them.

Killing cattle: success rate

Although dingoes are obviously successful at killing cattle, there are few data on how successful they can be. In the Fortescue River region of north-west Australia, however, researchers recorded only 73 interactions between cattle and dingoes over 7 years. Most interactions (81%) involved cattle reacting defensively to dingoes, and 36% involved attack on calves. However, only four calves were killed, so the dingoes' success rate was at best 15% (4/26), or only 0.05% (4/73).

Hunting sheep

There are many anecdotes about dingoes' prowess at killing sheep, none better than the assertion of one farmer in Queensland who believed a single dingo ripped out the throats of 600 sheep in one night; or perhaps a Victorian dogger's account of how a dingo killed 45 lambs in 60 minutes by running amongst panicking sheep, then grabbing a lamb by its nose and propping so that the lamb flipped over and broke its neck.

The only detailed evidence, however, is from Thomson's study in the Fortescue River region in north-west Australia. From a total of 61 attacks, 26 sheep (43%) were seriously injured, but only 8 (13%) were killed outright. Dingoes can easily outpace sheep but in most of these cases (42, 69%), they simply broke off the attack, and in some cases (19, 31%) attacked another sheep. Apart from running, sheep showed no defensive behaviour, and on 11 occasions (18%) struggled so little that dingoes were able to begin eating them without inflicting a killing bite.

Hunting rabbits

Dingoes usually hunt alone to catch rabbits, often by sighting and running them down, or by scenting and pouncing on a rabbit in a grass hutch. However, their greatest success comes from knowing where to find rabbit warrens, especially warrens containing rabbit kittens. Dingoes, in groups or alone, include those warrens in their regular hunting circuit, as indicated by the following example of a lone female dingo hunting near the upper Hale River in central Australia, where rabbits abound.

She sniffed at the entrances of several burrows then purposely proceeded straight along the river flat to the next group of warrens, about 200 m further. As she arrived, several adult rabbits who had been watching her quickly ran down holes. The female walked over the warren, very slowly, sniffing at most of the entrances, then stopped at one. She momentarily froze (front feet together, leaning slightly back on hindquarters, tail low, ears upright and forward), all the while looking intently down the hole, then she pounced down it. But she was unsuccessful and continued sniffing at the warren, revisiting some of the entrances. At one of them she poised as before, for 3 minutes, then pounced and emerged with a rabbit kitten in her mouth. She shook it vigorously and ate it completely, head first, in only 90 s. This pattern of hunting continued so that after an hour she had visited four sets of warrens, sniffed at about 80 entrances, and made four attempts to catch rabbit kittens, two of which were successful.

On another occasion a lone adult male walking along a creek bed suddenly rushed towards an adult rabbit sitting on top of a burrow about 20 m away. Surprisingly, the rabbit did not go down the hole but ran across the warren and up a steep hill with the dingo in hot pursuit and gaining. The rabbit suddenly doubled back to the warren around a fallen branch, and almost escaped down a hole. The dingo, however, grabbed it on its way down the hole, shook it, then retired to a nearby shrub and devoured it in 7.5 minutes. He began by pulling off and spitting out some fur from the rabbit's belly and then opened its body cavity with his incisor teeth and forepaws; eventually all that remained was some belly fur and about 1 m of large intestine.

Hunting birds

Many adult birds are captured when they are moulting (and thus unable to fly), otherwise dingoes eat young nestlings or newly fledged birds that are easily captured. For example, at a coastal salt lake in south-east Australia, a dingo 'cornered' six moulting swans in a small shallow inlet. When the swans made a break for open water, a second dingo rushed out of the fringing scrub and they swiftly dispatched all six swans with bites across the back behind the wings; but only four of the swans were eventually eaten.

In the coastal wetlands of northern Australia, fledgling magpie geese form a major part of the dingo's diet, and are captured in much the same way swans are. However, healthy and fully mobile adult geese are also obtained by stealing the kills of large raptors (predatory birds). This remarkable performance of unsuspecting co-operation and piracy between dingoes and large raptors is illustrated by the following example of a lone male adult dingo hunting on the floodplains of the South Alligator River in Kakadu National Park.

The dingo approached a flock of several hundred geese that were feeding on the ground, and from about 100 m began to skirt them slowly. The geese were soon alert. With heads upright they closely watched the dingo, but none flew away. The dingo continued prowling around them, sometimes giving the impression that he was going to rush at them, despite the large distance he would have had to cover. Suddenly an adult white-breasted sea eagle plummeted down and killed an adult goose. The geese had apparently been so distracted by the dingo that they had relaxed their vigilance against the aerial predator. The dingo immediately ran over, drove the eagle from its kill and ate the goose.

The eagle did not fly off with the goose because eagles are seldom capable of lifting off with an adult goose (up to 3 kg) and have to feed on the ground at the capture site. Dingoes similarly pirate the kills of swamp harriers and spotted harriers which are also common raptors on the floodplains of Kakadu.

Hunting rodents and other small prey

Dingoes usually hunt alone to catch small prey and mainly rely on their hearing and smell to find rodents, grasshoppers and other small prey moving about in grass and other vegetation. They capture these prey by pouncing on them with their forepaws. The posture of poising and pouncing is similar to that described above for hunting rabbit kittens in burrows.

Scavenging

Dingoes readily scavenge food, particularly cattle carrion which becomes plentiful during drought in arid regions of Australia. They usually gorge themselves on the carcasses of large prey either in one feed or intermittently over several days until it is all eaten. Sometimes rival dingo groups dispute over a windfall carcass, but possession ('nine-tenths of the law') prevails, so that otherwise subordinate groups can benefit from scavenging.

Dingoes seem to know where to find food in their living areas and regularly investigate these sources during hunting sessions. This knowledge was apparent in the example of dingoes in central Australia focusing on warrens containing rabbit kittens. Similarly, dingoes living in coastal regions regularly patrol the beaches and scavenge fish, seals, penguins and other birds that are frequently washed up. In Kakadu National Park, dingoes regularly scavenge the remains of prey underneath the nests and feeding platforms of white-breasted sea eagles.

DIETARY OVERLAP AND COMPETITION WITH FOXES AND FERAL CATS

Table 7.2 indicates the broad diets of dingoes, foxes and cats living in different habitats in Australia. To validate comparisons the data were derived from studies that either concurrently sampled all predators in each habitat or sampled each predator in different years in the same locality; thus there were four studies on the eastern highlands and one each on other habitats.

Mammals predominated in the diet of all three predators everywhere, but only dingoes ate large mammals. Instances of large mammals in the diets of cats and foxes were attributed

Table 7.2

The diets (% occurrence) of dingoes, foxes and cats compared between different habitats in Australia.

There are no foxes and rabbits in the northern tropics and the Barkly Tableland. Stock comprise cattle, buffaloes, horses, pigs and sheep. Other diet items include crustaceans, molluscs and fish. For definitions of mammal size see Box 7.1

	South-east Highlands			Central Desert			Barkly Tableland		Northern Tropics	
	Dingo	Fox	Cat	Dingo	Fox	Cat	Dingo	Cat	Dingo	Cat
Indigenous small mammal	10	26	24	18	42	81	65	86	34	86
Indigenous medium mammal	27	50	42	-	-	-	4	-	27	12
Indigenous large mammal	50	25	1	15	-	-	3	-	2	-
Introduced small mammal	2	2	6	8	<1	4	1	-	-	-
Introduced rabbit	13	8	6	56	82	55	-	-	-	-
Introduced stock	7	4	-	18	-	-	26	-	11	-
Bird	9	10	16	4	22	17	9	14	34	8
Reptile	3	3	12	12	11	55	12	29	<1	2
Insect	22	21	16	2	41	44	7	29	1	4
Vegetation	27	27	9	<1	9	4	-	-	7	20
Other	9	7	5	-	2	-	3	-	<1	-
No. of faeces or stomach contents	1635	1550	101	285	44	38	365	7	6722	49

to scavenging. Both dingoes and foxes tended to focus on medium-sized mammals, and cats focused on small mammals. Consumption data showed no clear distinction between predators' preferences for indigenous or introduced small and medium-sized mammals; rabbits were favoured by all three predators.

Foxes and cats ate more birds and reptiles than dingoes did, except in the coastal tropics where dingoes tended to focus on birds (magpie geese). Except in the eastern highlands, foxes and cats ate more insects and vegetation; they also included a broader range of other items than dingoes did (see Table 7.2).

In summary, there is considerable dietary overlap but dingoes tend to focus on larger mammals whereas foxes and cats focus on smaller mammals, reptiles and birds. In good seasons there is probably minimal competition, but in drought or after severe wildfires, food sources are dispersed and clumped, and competition for them is probably intense. At such times dingoes obviously survive the best, considering the drastic reductions in fox and cat numbers (see Chapter 9) and the inclusion of cats in dingo diet in all regions of Australia (Appendix C).

CHAPTER 8

POPULATION DYNAMICS

This chapter describes the major causes of mortality in dingoes and how populations may be limited by disease and social behaviour. Changes in limitation factors since European settlement have resulted in a massive increase in dingo numbers despite human control measures.

PARASITES, DISEASES AND ASSOCIATED CAUSES OF DEATH

About 38 species of parasites and pathogens in dingoes have so far been recorded (Appendix D). In Australia at least another 50 infectious organisms in domestic dogs have been recorded, all of which could become established in wild dingo populations.

The debilitating effects on dingoes are unknown for 45% of dingo parasites and pathogens but 29% (11/38) are known to be fatal, especially in pups (7/11 fatal infections). Lungworm, whipworm, hepatitis, coccidiosis (2 species), lice and ticks can kill pups and thereby decrease recruitment to dingo populations. Hookworm (2 species), heartworm and distemper have killed adult dingoes. The latter two organisms have virtually eliminated entire dingo populations, as described below.

Canine distemper

Cattlemen of long standing in the Northern Territory sometimes remark on a sickness which occasionally afflicts dingoes so badly that it wipes out local populations for several years. Most recall that the biggest epidemic accompanied the widespread, severe drought of the early 1960s, and all describe similar symptoms. Dingoes had shrunken, runny eyes, and most seemed to have lost their sense of smell, hearing and sight so that they were easily killed with sticks.

In 1970 a research team visited Brunette Downs on the Barkly Tableland to investigate reports of dingoes staggering about and dying around watering points. A sample of 18 dingoes was collected in two days; some were so sick that they could be approached and shot at a range of only 1–2 m. Post-mortem indicated that most had distemper. At the time, the Barkly Tableland was in drought, but this had followed a run of exceptionally flush years and a plague of long-haired rats for about 3 years. The dingo population had accordingly flourished, but when that abundant food source suddenly disappeared in 1970, the dingoes reached starvation point. This was indicated by the empty stomachs of all afflicted dingoes and many apparently healthy dingoes; others were eating the carcasses of cattle that had also succumbed to the drought.

According to reports of sick and dying dingoes from government stock inspectors and cattle station lessees, the distemper epizootic spread right across the Barkly Tableland and south to the Petermann Ranges in the west and the Simpson Desert in the east. It is interesting that long-haired rats were also recorded, sometimes in plague numbers, in localities where afflicted dingoes were found, even as far south as the Simpson Desert. Long-haired rat plagues are rarely recorded, about 1 in 9 years on the Barkly Tableland and even more rarely in the Simpson Desert; perhaps there is a connection?

The impact of the distemper epizootic on dingo numbers was catastrophic. Before the epizootic, the same research team used to collect bimonthly samples of 30–50 dingoes on the Barkly Tableland, and a famous dogger, Syd Ballard, regularly scalped over 1,000 dingoes each year. By 1971 the drought had broken but very few tracks or other dingo signs were seen and only four dingoes were sampled by the research team in 20 days. The scarcity of dingoes was verified by local cattle station lessees, who also told the research team that many of their cattle dogs had also died; and that the dogger had gone fencing. Most likely, the anecdotes of the old cattlemen referred to similar distemper epizootics.

Heartworm

In the wet-dry tropics of northern Australia dingo mortality from heartworm is probably another form of widespread but intermittent population control; however, its epidemiology (origin and pattern of spread through host populations) is not completely understood. Heartworm is mostly spread by mosquitoes, perhaps restricted to several species. Most domestic dogs become infected and virtually all untreated adults eventually die. Most dingoes (73%) in Kakadu National Park are infected by a few to over 200 adult heartworms per dingo. Sometimes there is high mortality. For example, at Kapalga, there have been two epizootics of heartworm during the past 14 years, each wiping out about half the known dingo population (see Plate 21). Given the high mortality rate for domestic dogs and the apparently high mosquito populations in the bush, it is perhaps surprising that the dingo mortality rate is not greater. Some clues can be gained from the circumstances associated with the Kapalga epizootics, as follows.

Five factors common to the years of the heartworm epizootics were:

1 The preceding wet season began very late (January/February) and rainfall was well below average. This particular weather pattern may have encouraged high populations of the mosquito species that transmit heartworm most readily, so that more dingoes were bitten (and infected) than usual.
2 Most of the dingoes that died lived in floodplain not forest habitats.
3 Deaths occurred in August and September, about 6 months after the preceding wet season had begun. This period is consistent with the time required for enough heartworm larvae to grow into adults and lodge in the heart and pulmonary artery to cause circulatory problems.
4 There were very high dusky rat populations, but despite this abundance of easily caught food, dingoes apparently still died from starvation. For example, three healthy radio-collared dingoes died in just 2 weeks; they just couldn't run and seemed exhausted whenever they tried.
5 No pups were affected, probably because most were born in May and June so they were not exposed to high vector (mosquito) populations.

These observations support the notion that specific mosquito species transmit heartworm and require specific breeding conditions to attain high numbers.

After both heartworm epizootics at Kapalga, the dingo population recovered quickly with a higher than usual recruitment of pups in the following year. Some subordinate females' pups may have been reared and recruited in addition to those of dominant females.

Mange

Sarcoptic mange is probably the most widespread parasitic disease in dingo populations throughout Australia, but it is seldom debilitating. The highest incidence (20%, Appendix D) was recorded in the Fortescue River region of north-west Australia where 21% of mange-affected adult dingoes were in poor condition compared to 5% for mange-free dingoes; however, only one dingo was suspected to have died from mange. Males are twice as likely to be affected by mange as females are. It is likely that mange is symptomatic of other diseases, or associated with particular prey species. For example, at Kapalga in northern Australia high incidences of sarcoptic mange are nearly always associated with plagues of dusky rats.

Relationships between parasites and diet

It seems there are close interrelationships (cycles) between some parasites and the major prey eaten, and such relationships are habitat-specific, as indicated by the following examples. In eastern Australia the hydatid tapeworm, *Echinococcus granulosus*, is associated with a sylvatic (macropodid-dingo) cycle and a domestic (sheep-dog) cycle. In central Australia dingoes mostly eat rabbits and consequently have relatively high infestations of rabbit stickfast fleas and the tapeworm *Taenia pisiformis*. By comparison, the most common parasites of dingoes in the nearby Barkly Tableland are lice and another, unidentified, tapeworm species which are probably derived from eating long-haired rats, a major diet item in this region. Finally, as indicated above, sarcoptic mange is usually associated with plagues of dusky rats in the wet-dry tropics.

RABIES AND OTHER POTENTIAL EXOTIC DISEASES OF DINGOES IN AUSTRALIA

More than two millennia have passed since Acteon the hunter came across the goddess Diana and her nymphs while they bathed. The chroniclers of ancient Greece recorded that the divine punishment for his voyeurism was to be savaged by his own hounds, magically made rabid at Diana's command. Fear of rabies, or hydrophobia, has since reached awesome proportions in human minds. Rabies has horrifying symptoms in humans, and once symptoms appear death is inevitable. About 15,000 human deaths and immense economic losses of livestock are incurred each year.

All warm-blooded animals are susceptible to rabies, though some species are more susceptible than others, and some spread the virus more readily than others do. Dogs (and presumably dingoes) are only moderately susceptible; wolves, jackals, coyotes and foxes are very highly susceptible.

In general, animals with rabies show one of two forms: furious (excited) or dumb (paralytic) rabies. Both forms are infectious and the virus is usually transmitted in saliva, but sometimes in urine. For dogs, the incubation period (between infection and appearance of symptoms) is generally 2–6 weeks, sometimes 6 months; for humans it takes 6–12 weeks, sometimes more than 12 months.

Rabies is virtually pandemic, absent only in Australia, New Zealand, Papua New Guinea, Fiji and other islands in Oceania, the United Kingdom, and Antarctica. Although rabies is endemic in Asia, inadequate reporting makes it difficult to determine its distribution and extent. For example, in 1979 India reported 72 rabid dogs, Vietnam reported 33,145, and Indonesia, Thailand and the Philippines did not report any; yet post-attack anti-rabic treatment was given in all countries. At 36 deaths per million inhabitants, India has the highest annual rate of human fatality in the world.

Northern Australia's proximity to Asia provides many opportunities for the introduction of exotic diseases. Darwin has an international airport with many arrivals from Asia. Cargo shipping arrives directly from Asian ports, themselves infested with rats and other vermin that can jump ship. Indonesian fishermen and other Asian boat people regularly land on Australia's northern shores, sometimes undetected, and some are reputed to bring dogs and other livestock. Itinerant yachts carrying pet dogs and cats frequently commute between Darwin and Indonesia. South-east Asia has at least eight diseases that could be carried by dingoes (Table 8.1), and some of them could be disastrous for people (e.g. rabies) or livestock (e.g. surra).

It is interesting, perhaps surprising, that rabies is not present in Australia, given its presence in nearby Indonesia and the long history of Asian seafarers visiting Australia with their dogs and other animals. Perhaps rabies has been introduced, on many occasions, but not become established. The epidemiology of rabies is not fully understood. Our present knowledge depends heavily on the ecology of the species involved in its propagation, and often involves a sylvatic cycle (wild animal vectors rather than domestic animals). However, data on the transmission of rabies by marsupials are scarce. Rabies and other exotic diseases, once introduced, could become established in wild and feral animal populations in coastal northern Australia without being detected because of isolation or lack of clinical signs. Consequently, such diseases could spread southwards to susceptible livestock and to areas of high human density. More study is necessary to understand intraspecific and interspecific interactions between potential vectors of rabies and other exotic diseases.

Table 8.1
Diseases in Asia which could be hosted by dingoes and other animals in northern Australia

Disease	SE Asian distribution	Natural host
Aujeszky's disease	Mainland & islands	Pig, dog, cat, rats, cattle
Glanders	Mainland & islands	Equids, dog, cat, humans
Japanese encephalitis	All west of Bali	Waterbirds, reptiles, mammals, amphibia
Rabies	All except PNG Singapore, most of Malaysia, and parts of Indonesia	Mammals
Screw-worm fly	Mainland & islands	Mammals
Surra	Mainland & islands	Livestock, pig, dog, cat
Transmissible gastroenteritis & porcine epidemic diarrhoea	Mainland & most islands. No information from Indonesia	Pig, dog, cat
Tropical canine pancytopaenia	Mainland	Dog

POISONS, TRAPS, GUNS AND OTHER CAUSES OF DEATH

Dingoes have been poisoned, trapped and shot for over 200 years in Australia, and enormous sums of money and effort have been spent in the effort. But to what effect? Although evidence is scarce, it seems that poison campaigns can succeed in arid regions but not so well in moister habitats. Trapping is more efficient than poisoning but is usually much less economical.

Broadscale poison campaigns: variable success

In non-pastoral areas of semi-arid north-west Australia, two aerial broadcast bait campaigns killed 90% and 44% of radio-collared dingoes respectively. In arid pastoral areas, similarly conducted campaigns killed an average 62% of dingoes, which indicates that all these campaigns were quite successful.

However, in arid central Australia contrasting results have been obtained. An extensive aerial campaign that broadcast strychnine baits was a total failure: the numbers of dingoes subsequently increased! It failed because the fat baits were unpalatable and because there was plenty of food in the prevailing flush conditions. Soon after, drought prevailed in the same areas and another campaign was conducted with baits loaded with either strychnine or 1080 poison. This campaign was very successful: about 70% of the dingo population was killed irrespective of the poison used. Its success was due to attractive fresh meat baits laid along established dingo trails and watering points, and the food shortage imposed by the drought.

In the Victorian Highlands, six extensive aerial broadcast bait campaigns were conducted between 1953 and 1970 involving about 200,000 strychnine and 1080 poison baits. All were apparently unsuccessful since over the same period the number of bounties paid out on dingoes remained fairly constant (about 800 annually).

However, more recently in the south-east highlands next to Kosciusko National Park, ground bait programs killed 22% of radio-collared dingoes; leg-hold traps captured 56% of the known dingo population. The success of the poison program was limited by the rapid loss of toxicity from the baits (1080 poison) after their distribution, their rapid removal by non-target animals, and their unattractiveness to dingoes, many of which apparently preferred to kill natural prey.

Trapping and shooting

There are few data on trapping success. In the 1960s there were about 26 professional doggers in north-east Victoria. Each set 25–175 traps in lines up to 80 km along mountain tracks and animal trails. In a sample involving 12 doggers over 6 years, only 13 dingoes were captured in 4,796 trap-nights (369 trap-nights/dingo).

During the same period and in the same areas, more careful placement of traps on fresh dingo signs only increased the success rate to 92 trap-nights/dingo (1,108 trap-nights, 12 dingoes). More recently, similar methods achieved a success rate of 60 trap-nights/dingo near Kosciusko National Park. Clearly, trapping is time-consuming but perhaps more economical than baiting.

Shooting drives are rarely conducted these days. They are claimed to work well for eliminating foxes, but not for dingoes. For example, it took 40 horsemen, 20 guns and 2 days to shoot 1 dingo during a drive in New South Wales in 1966.

Dispersal sinks

One of the main, consistent causes of dingo mortality is a cycle involving dingo population density, food supply and human control. When food becomes scarce for a large population of dingoes in a 'safe' area they move away, singly or in groups, to resource-rich, 'vacant' pastoral and agricultural areas. These are 'danger' areas, where human control measures are intense. Dingoes are poisoned, trapped or shot there, so that 'sinks' (vacant areas) are created and the dispersal-mortality cycle is perpetuated.

For example, over 9 years in the Fortescue River region of north-west Australia, 25 lone dingoes and 6 groups (comprising 24 individuals from fractured packs) dispersed from non-pastoral 'safe' areas to sheep paddocks where control was intensive. All except one of these 49 dingoes were eventually killed — either trapped, poisoned or shot.

Other mortalities

Other, less common, sources of dingo mortality include: collision with motor vehicles; being gored by buffalo; being kicked by cattle; drowning by kangaroos; snake-bite; predation of pups by wedge-tailed eagles. In Asia, predation by humans probably account for most dingo deaths. In the 'dog' abattoirs in north-east Thailand, for example, about 200 dingoes are butchered each week for human consumption.

SOCIAL FACTORS LIMITING POPULATION GROWTH

Reproduction and social interactions in a pack of captive dingoes, Curly's Mob, were described in Chapter 6. Over the 3-year span of that study, the potential yearly increases in dingo numbers were 7, 18 and at least 43, respectively (Figure 8.1). In the latter two years these numbers were not realised due to infanticide and the killing of two adult females, so that at the end of the experiment only 30% of the pups had survived and the total number of dingoes in the enclosure was 16, less than half the potential increase (Figure 8.1).

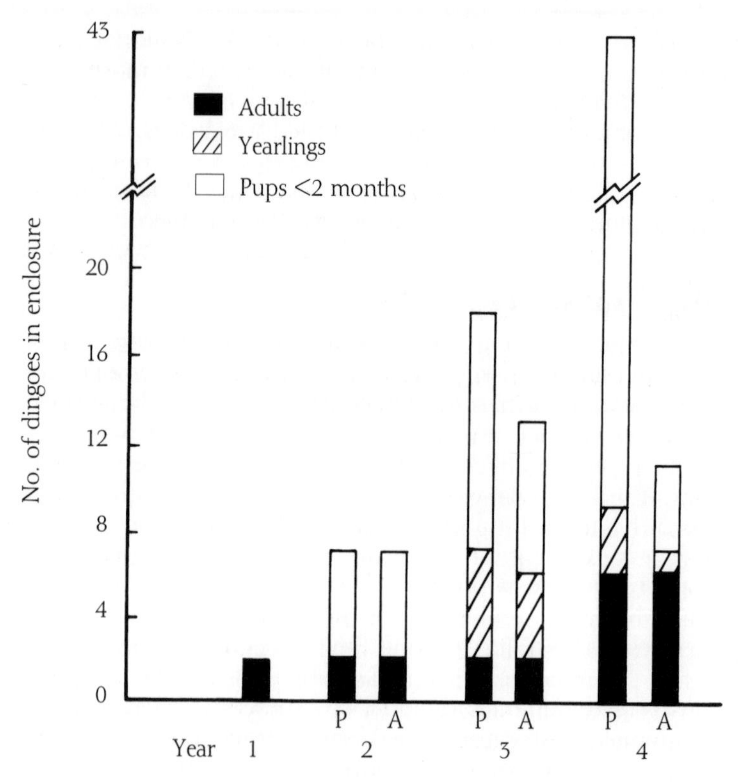

Figure 8.1

The potential (P) and actual (A) increase in the size of a dingo pack in captivity over 3 years; the difference is due to mortality. Most deaths were due to infanticide: the founding and dominant female killed and ate all her daughter's pups; 2 yearling females were killed by a subgroup of adults and yearlings.

From other data (albeit less precise) one can draw similar conclusions about wild dingo packs in arid (Simpson Desert), semi-arid (Fortescue River) and tropical (Kakadu National Park) regions of Australia. In a study of captive wolves in Europe, only 14 pups out of a possible 90 survived over 5 years, representing only 16% of the potential increase in numbers. These examples suggest that social factors may limit population growth, although these factors may vary between species. The main limitation controls are the inhibition of breeding, mating preferences and infanticide; these are described below.

Breeding suppression and mating preferences

Most wolves suppress their breeding by preventing subordinates' copulation. That is usually achieved by the alpha female: she blocks subordinate females' social and sexual activities with threats, intense attacks, and imposition behaviour to drive them away from males. The alpha male also suppresses subordinate males' courtship activities, but less frequently and less intensely. Usually male wolves are more sexually interested in the alpha female than in subordinate pack females, even those displaying vulval bleeding; and males accepted by the alpha female aggressively prevent other males from approaching her. In some cases, such behaviour has also been attributed to a litter-mate incest barrier.

In a study of captive wolves, females on reaching sexual maturity established and maintained a preference for a certain alpha male. Another study indicated that some males show a preference for their own mother irrespective of her rank.

Other studies of captive wolves have suggested that aggression by alpha wolves may interfere with the production of hormones in both sexes so that subordinate females have less sex appeal, the onset of their first ovulation is delayed, and the copulatory urge to subordinate males is diminished. Such endocrine function may also partly explain the incest barriers attributed to wolf packs that have only one pair of parents reproducing. The inhibition of sexual behaviour by a dominant individual is apparently widespread in the animal kingdom and has been reported, for example, in baboons, dwarf mongooses, buffalo and male sheep.

There is little known about breeding suppression and mating preferences in dingoes. High-ranking males and females in Curly's Mob sometimes aggressively suppressed the sexual soliciting and mating activities of low-ranking dingoes, especially the alpha male (Curly) who inhibited all the five subordinate males' mating attempts at the alpha female (Toots) and the beta female (Genevieve), except for some occasions when he was in copulatory tie with Toots. When the younger females came into oestrus (after Toots and Genevieve) Curly showed less sexual interest in them and eventually allowed the most persistent male (Beethoven) to mate with them.

Some, but not all, of Curly's Mob showed mating preferences. The most obvious were the alpha pair's preference for each other, the beta female's preference for the alpha male, and the low-ranking Cleopatra's preference for Beethoven (who was usually high-ranking). All other males and females solicited and attempted to mate indiscriminately. There were no litter-mate incest barriers in evidence since several litters resulted from matings between brothers and sisters.

Infanticide

Although the mating attempts by some low-ranking members

of Curly's Mob were suppressed, all oestrous females copulated and became pregnant. The dingoes' main method of suppressing breeding was infanticide whereby the alpha female killed all the pups of all other females.

Infanticide is widespread among carnivores and occurs in several contexts. Infanticide by males is common practice with lions, less so with tigers, pumas, cheetahs, brown bears, polar bears and coatis. It is suspected of domestic and feral house cats, dwarf mongooses and spotted hyenas. The practice of one breeding group killing the young of another group (extra-group infanticide) is carried out by coyotes, golden jackals, African hunting dogs and female lions. Sometimes mothers abandon their single-offspring litters, as in lions and grizzly bears (intra-group infanticide); but infanticide by females other than the mother is more common and occurs with African wild dogs, captive wolves, captive red foxes, brown hyenas and dwarf mongooses. In all such cases, infanticide was committed by the dominant female who killed or otherwise caused the death of subordinate females' offspring, as was the case for Curly's Mob. In all cases, except for brown hyenas, the dominant female, the 'exploited' subordinate females, and the killed offspring were all close relatives, and in most cases mothers killed their daughter's offspring.

An important consequence of dominant female infanticide (with the exception of wolves) is that subordinate females help to rear and even suckle the offspring of the dominant female. If all subordinate female dingoes had whelped in the final year of the Alice Springs study (Curly's Mob), there might have been more lactating females (n=6) than surviving pups (n=4), and they might have been able to continue suckling long after their own mother's supply had dried up. This consequence has further implications for dingoes in Australia, as follows.

THE EVOLUTION OF DOMINANT FEMALE INFANTICIDE BY AUSTRALIAN DINGOES

Dominant female infanticide appears to be a reproductive strategy for making sure that when the dominant female is breeding, her offspring not only receive additional care from subordinate helpers, but have no competition for resources from other females' offspring.

Dominant female infanticide may be an evolutionary consequence of an ecological fluctuation or a change of hunting style which in turn changes the social structure of a species. Thus there are likely to be conditions that provoke conflict between females of the same group over whether subordinate females also breed or only help rear the offspring of the dominant female. If this is the case, dominant female infanticide may be just a (final) extension of a behaviour pattern that suppresses subordinates' mating. The evidence for dingoes suggests this is so, but it still leaves the question of why subordinate females waste energy in breeding if their offspring are to be killed.

Dingoes are primitive canids widely distributed across Asia and Australia. Their social organisation and reproductive biology still closely resemble that of their wolf forebears despite their isolation in Australia for about 4,000 years. Most of the Australian mainland, including the northern 'wet-dry' tropics, is intermittently or regularly afflicted with severe drought so that most native mammalian species must adapt to survive dry periods when resources may be extremely limited.

In pre-European times most prey species and free-watering points for dingoes would have become extremely scarce during drought, so that they probably faced starvation. Under such conditions, the mortality rate was probably high, particularly for pups, including the offspring of the alpha pair. The low food supply during droughts would also have made packs split into smaller units. Such a capricious environment may have demanded the common reproductive selection strategy: the more pups born, the greater the chance that some would survive adverse periods. Since most breeding dingoes would have been closely related, at least some of the alpha pair's genes would survive to the next generation if all pregnancies went to term and if some of the smaller pack units were able to survive drought in the sparse resource points of that period. This may also explain why the dingo alpha male does not always suppress the breeding activities of subordinate males (as alpha male wolves do), nor kill their offspring.

Evidence supporting this hypothesis is provided by the inverse relationship between the pack size and habitat quality of dingoes in the Fortescue River region of north-west Australia. It has also been reported that the litter and pack size of wolves in Minnesota were inversely related in areas where food supplies were low.

NATURAL AND UNNATURAL FACTORS AFFECTING THE GROWTH OF DINGO POPULATIONS IN AUSTRALIA

After dingoes came to Australia about 4,000 years ago, their numbers would have been kept low by natural influences — social factors and diseases would have systematically and periodically reduced them (Figure 8.2).

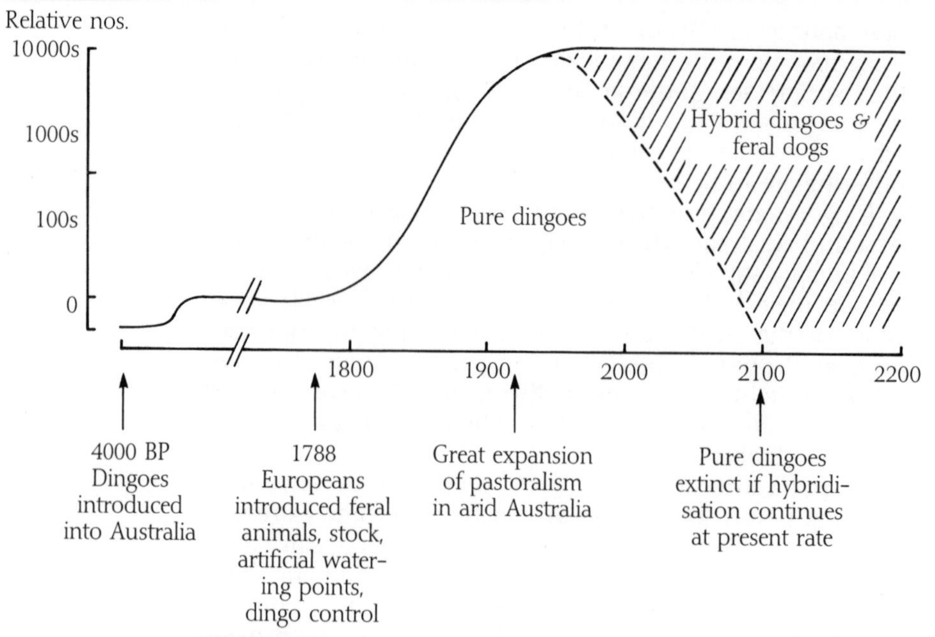

Figure 8.2

The rise and fall of pure dingoes in Australia. After Asian seafarers brought them here c. 4,000 years ago their numbers remained low until European settlement changed their conditions and greatly increased their food and water, particularly during droughts, so that dingo numbers probably trebled by the mid 20th C. Since then the numbers of wild canids living in the bush have remained high but the proportion of pure dingoes is rapidly declining due to hybridisation with domestic dogs. If the present trend continues, pure dingoes will be extinct by the end of the 21st C.

With European settlement and the development of the pastoral industry in inland Australia about 100 years ago, dingo numbers exploded as food became more plentiful (rabbits, stock, some native macropodids) and artesian bores and dams increased water supplies during drought. Also, control methods for dingoes (trapping, poisoning, shooting) are not fully efficient and in many cases probably only fracture packs into smaller units, each with a breeding female. In this way the dingo's population suppression method, dominant female infanticide, is discouraged, so that dingo numbers increase abnormally during flush periods.

This process is illustrated by wild wolf packs in Alaska. Hunting split their packs into smaller units, and this interfered with their natural breeding controls. The result was high breeding rates, with 89% of adult females pregnant and birth rates at 44–60%.

Most of the victims of many dingo control programs were probably young and old dingoes, so that surviving populations tended to consist mainly of middle-aged dingoes; these animals were probably the most likely to mate and raise litters successfully by themselves.

The total dingo population in Australia probably peaked between 1930 and late 1950s — soon after rabbits had become well established in two-thirds of the mainland, and after the pastoral industry had begun to benefit greatly from underground water. Since then, so-called dingo populations have remained high, but as Figure 8.2 shows, these populations consist of pure dingoes, hybrids and feral domestic dogs. The decline of pure dingoes is discussed in Chapter 10.

An indication of the magnitude of the increase in the dingo population can be gauged by comparing the reports of the early explorers of central Australia with those of more recent adventurers. For example, McDouall Stuart and members of his party recorded that they rarely saw or heard dingoes despite crossing the 'centre' on three occasions between 1860 and 1862. The first zoologists, members of the Horn expedition in 1894, simply stated: 'the dingo is fairly numerous' but their only indications of 'abundance' were that they found five drowned dingoes in an Aboriginal well near Lake Amadeus, and that a dingo stole a hare-wallaby skull from their camp one night.

Now, 100 years later, dingoes abound in the 'centre' and it is difficult to imagine anyone not recording encounters with them. In the early 1960s the famous dogging team, Peter Allen and his sons, regularly claimed the scalps of up to 4,000 dingoes annually, taken along the foothills to the Macdonnell Ranges. Even inexperienced scientists were able to trap 30–50 dingoes each month over several years in central Australia in the 1960s and 1970s.

Recent radio-tracking and mark-recapture studies have indicated dingo densities of 1/6 km^2 in Kosciusko National Park and 1/7 km^2 at Kapalga in Kakadu National Park. In the Fortescue River region the density of dingoes in pack territories varied between 1/4 km^2 and 1/29 km^2 (mean 1/12 km^2).

CHAPTER 9

PREDATOR-PREY INTERACTIONS

This chapter examines the nature and consequences of interactions between dingoes and their prey and between dingoes and their native competitors. It also asks: what role have dingoes played in shaping and maintaining ecological relationships in Australia?

New arrivals to a continent must, to some extent, affect native species by preying on them or competing for resources, particularly those with similar living requirements (ecological niches). This is true of dingoes in Australia but some of their interactions have been mediated by the European exploitation of Australia, so that there are three broad types of ecosystems to consider, as follows.

1 *Natural ecosystems that preceded European settlement.* Although there were relatively few dingoes they were new eutherian (placental) players in a land of marsupials and they introduced new predation and competition rules.

2 *Relatively undisturbed ecosystems (pristine wilderness areas) after European settlement.* Dingoes continue to live here in packs, and their populations appear to be regulated by natural processes.

3 *Ecosystems disturbed by pastoralism and other aspects of European settlement.* Not only have dingo numbers greatly increased but changes to their social structure and reliance on introduced prey has had dire consequences for some native fauna and for pastoralists.

DID COMPETITION FROM DINGOES CAUSE THE EXTINCTION OF THYLACINES AND TASMANIAN DEVILS ON THE AUSTRALIAN MAINLAND?

The thylacine, also known as the Tasmanian wolf, is a marsupial carnivore about the size of a dingo, and perhaps can be considered as the evolutionary equivalent to the eutherian dingo. According to fossils and Aboriginal rock paintings, the thylacine was once distributed throughout Australia, but 'suddenly' disappeared from the mainland about 3,000 years ago. Dingoes had arrived in northern Australia about 500 years earlier and probably spread across the continent fairly quickly; the oldest dingo fossils, in southern Australia, are dated at 3,450 years BP. The demise of the thylacine is often attributed to the arrival of the dingo, but until now no convincing arguments about the competition between them have come forward. How did dingoes overcome thylacines? Two hypotheses are presented, as follows.

The superior adaptability hypothesis

This hypothesis hinges on the superior social organisation of dingoes during critical periods when food supplies were scarce, widely dispersed and clumped during drought or after extensive wildfire. Only dingoes could form large integrated packs and cooperate to catch large prey, and to defend carcasses, water and other crucial resources. Thylacines, on the other hand, hunted alone or in pairs, and could not compete against the weight of dingo numbers during those critical periods.

What is the evidence? Thylacines are recorded as having a stiff gait; they could not run after their prey (mainly macropodids) as fast as dingoes could. They apparently located their prey by scent and tired it by dogged pursuit, alone or in pairs. There are no records or anecdotes of thylacines hunting cooperatively in groups. This apparent lack of a group system is supported by bushmen's observations that the thylacine was normally mute except for a coughing bark. Social animals, like dingoes, have a large vocal repertoire for communicating a variety of functions over large and short distances.

Dingoes are now presenting foxes and feral cats with a similar kind of competition. The most common prey species of dingoes, foxes and feral cats in central Australia are rabbits and small rodents. During a drought between 1969 and 1972 these prey became scarce and dingoes changed their diet to red kangaroos and cattle carcasses. Dingoes were more successful at catching kangaroos by hunting cooperatively than alone, and stable packs of dingoes defended carcasses and waters more successfully than less cohesive groups could.

During the first year of the drought, cats and foxes were also seen scavenging cattle and kangaroo carcasses, but sightings of them ceased and their tracks disappeared about midway through the drought. They most likely starved because many emaciated cats suddenly appeared around homesteads and 180 of them were killed by park rangers at Uluru (Ayers Rock) over a 6-week period in 1970. Dingoes probably contributed significantly to the demise of cats and foxes by their increasing monopoly of carcasses as the drought persisted; it was also observed that foxes avoided dingoes at shared resources and an increase of cat in the dingo's diet was recorded. In any event, there were no signs of cats or foxes until the drought broke about 2 years later, after the rabbit and rodent populations had resurged.

It is possible that even three millennia ago dingoes were not only more efficient hunters of large prey but excluded thylacines from crucial food resources. But, unlike today, there were no cattle or rabbits to bolster food supplies during drought and flush times respectively, which means that thylacine populations would have been less able to recover after the intense periods of competition. Similarly, in North America, the superior competitiveness of coyotes is believed to limit the distribution of red foxes.

The Tasmanian devil, another marsupial carnivore about half the size of a dingo, was also widespread throughout Australia about 4,000 years ago, but its population also declined and it became extinct on the mainland about 450 years ago. Its demise is also attributed to competition with dingoes, as indicated below.

Observations of devils in Tasmania, where they are still common, confirm that they hunt alone, and that although they can catch a variety of live prey they subsist mainly on carrion (macropodids and sheep). Up to 12 individuals have been recorded around carcasses, noisily squabbling for access and occasionally being physically aggressive. It seems that devils are essentially solitary animals and their aggregations around carcasses are not cohesive social units, so it is quite likely that devils could not successfully compete with dingoes when food was scarce during drought and after fires.

The disease hypothesis

The second hypothesis is that dingoes brought to Australia a pathogen that wiped out thylacines and devils on the mainland. Such a pathogen could have operated independently or synergistically with dingo competition. There is no direct evidence that dingoes introduced any disease into Australia, but it is possible since they probably transported a parasitic louse from Australia to Asia. It is also known that epizootics devastated the devil population in Tasmania in c. 1909 and 1950, and that the thylacine population was severely reduced between 1900 and 1920. This reduction is thought to be too rapid to be attributed entirely to human persecution, so the thylacine probably fell foul of the same pathogen that attacked the devils.

The organism responsible for these epizootics is unknown, but a strong contender is the disease known as toxoplasmosis which is caused by a protozoan parasite and was probably spread in Tasmania by feral cats. This disease is fatal to many marsupials, especially the Dasyurids. Since toxoplasmosis also occurs in dogs in Australia and Asia and thus is likely to occur in dingoes, it is possible that this pathogen or another, such as rabies, was introduced by dingoes and afflicted the mainland populations of thylacines and devils in pre-European times.

THE ROLE OF DINGO PREDATION IN THE DEMISE OF MEDIUM-SIZED MAMMALS IN CENTRAL AUSTRALIA

Much has been written about the demise of native fauna in central Australia and the role of predation by foxes and feral cats, as well as the role of competition from rabbits, cattle and sheep. Much has also been written about changes in land management practices, particularly the use of fire by Aborigines and pastoralists. Dingoes, however, probably have played as great, if not a greater, role in the decline and extinction of medium-sized mammals, especially species that sheltered on the surface amongst grass and shrubs.

In central Australia, before the 1930s, 14 species of bandicoots, macropodids and rat-kangaroos were reasonably common in the areas where cattle now graze, but only five species survive today (Table 9.1), and of these, two are rare and endangered.

The arrival of rabbits, foxes, feral cats and stock in central Australia

Rabbits had arrived in central Australia (Charlotte Waters) by 1896 but it seems that their numbers did not rapidly increase until about the 1930s, when foxes arrived. Foxes were not recorded by Finlayson during his extensive journeys in the early 1930s nor did he indicate that the local Aboriginal tribes (Luritja) had seen them. The presence of feral cats, however, had been recorded in central and northern Australia at the

Table 9.1

The status of 14 species of macropodids, bandicoots and rat-kangaroos in central Australia before and after about 1930.

Some increased, most declined. Although some of these mammals were already rare in the early part of this century, the greatest decline occurred during the 1930s and 1940s when the cattle industry greatly expanded once artesian water came into use. Many reasons for the decline have been suggested but none, until now, have considered dingo predation.

* = species that sheltered in hutches (squats) in long perennial grasses or under small shrubs.

+ = species still extant in other regions of Australia, notably on offshore islands.

	Before 1930	Today
Red Kangaroo	Common	Abundant
Euro	Common	Common
Black-footed Rock Wallaby	Common	Common
Greater Bilby	Common	Rare+
Rufous Hare-wallaby*	Common	Probably extinct+
Spectacled Hare-wallaby*	Common	Extinct+
Golden Bandicoot*	Common	Extinct+
Burrowing Bettong	Common	Extinct+
Lesser Bilby	Present	Extinct
Crescent Nailtail Wallaby*	Common	Extinct
Desert Rat-kangaroo*	Present	Extinct
Central Hare-wallaby*	Present	Extinct
Desert Bandicoot*	Present	Extinct
Pig-footed Bandicoot	Uncommon	Extinct

end of the 19th century. For example, the explorer Winnecke described a feral cat in the Simpson Desert in 1883 as 'a wild cat of extraordinary size; the brute was nearly as large as a leopard'. The feral cat may have been present even earlier, since it was the only non-indigenous mammal to which the Luritja tribe in central Australia gave a specific name. This suggests that the feral cat may have had a longer tenure in Australia than is usually supposed, and that it made its way into cental Australia long before Europeans arrived there. It may have migrated from the early eastern settlements, but was more likely deliberately introducted or survived ship-wrecks along the northern and north-west coasts. Finlayson indicated that cats were plentiful all over central Australia by the early 1930s. Presumably, however, feral cats, like foxes, increased as rabbit populations grew.

The first stock (4,300 sheep) were brought through central Australia in 1871 to the Roper River in the tropical north. The first stations were established in 1872 at Alice Springs (by E. M. Bagot) and at Owen Springs (by J. Gilbert), and by 1885 land applications were lodged for virtually all of central Australia. By 1888 there were 50,000 cattle, 10,000 sheep and 4,500 horses in central Australia; this rose to about 160,000 cattle in 1930 (Figure 9.1). Whereas sheep numbers declined

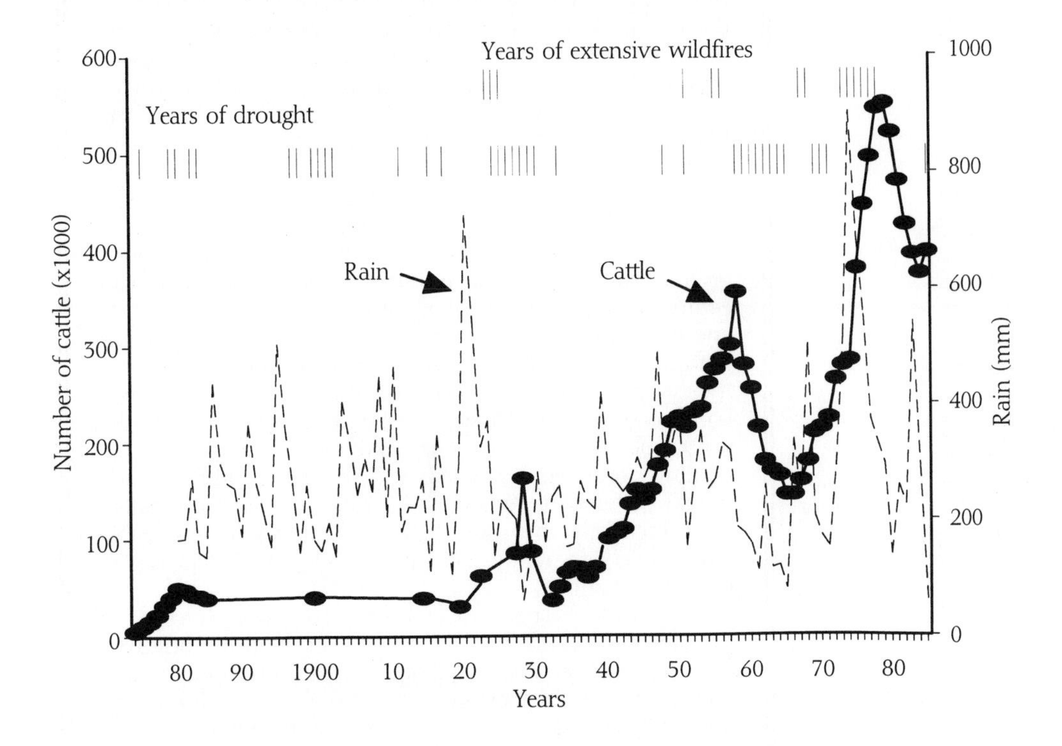

Figure 9.1

Cattle, droughts and wildfires in central Australia. In the 1930s sub-artesian bore water allowed the cattle industry to greatly expand, both in cattle numbers and grazing area. Although unpredictable, droughts are common in arid Australia. Yet cattle numbers quadrupled between the worst droughts on record (1924–30, 1958–65). Wildfires are also common in most habitats, but were rarely widespread in central Australia until Aborigines began to settle at mission stations and ceased their traditional small, cool burns. Less frequent but extensive, fierce fires and cattle grazing subsequently exposed many medium-sized native mammals to dingoes, whose numbers, paradoxically, were increased by the same conditions. Dingo predation was so great that some native species became extinct. In later droughts other species subjected to similar pressures including competition from rabbits and dwindling refuge areas were mopped up by feral foxes and cats, also increasing by then.

dramatically by 1920 and remained low thereafter, cattle numbers continued to increase despite the setbacks of droughts (Box 9.1).

Box 9.1
Droughts in central Australia: a prerequisite for the extinction of native mammals?

According to the Commonwealth Bureau of Meteorology, a drought is defined as a period of 10 months or greater in which no rain falls or rain is inconsequential and which overlaps the (summer) months of maximum vegetation growth. Since records commenced in central Australia in 1874 there have been 17 droughts up to 1992, averaging (± standard deviation) 22 ± 15 months. The most severe droughts, in terms of rainfall deficiency, were: 1925–30, 1958–68, 1951–53, 1931–36, 1904–6, 1899–1901 and 1911–16.

During drought rabbit and native rodent populations decline, but dingo numbers remain high because they can eat the carcasses of cattle that die from starvation. Feral cats and foxes decline because they cannot compete for carcasses with dingo packs. Before rabbits or cattle came to Australia the dingo's main prey were the medium-sized wallabies, bandicoots and rat-kangaroos that were common on the central plains. Although dingoes ate more rabbits and cattle as dingo numbers increased with the cattle industry, they would still have exerted considerable pressure on native prey, especially the species that became more vulnerable after cattle and rabbits ate their homes — the perennial grasses and shrubs.

The two longest and most severe droughts on record, 69 months in 1925–30 and 89 months in 1958–65, occurred over the cattle industry's greatest expansion period, which concurred with the increase in dingo populations. It is probably no coincidence that nine native mammals became extinct over the same period and a further two species became rare.

The expansion from the 1930s was due to the establishment of supplementary water from artesian bores, which also allowed cattle to graze further from natural water sources and survive longer in drought. After the most severe drought on record in central Australia (5.8 years), cattle numbers quadrupled from about 36,000 in 1932 to a peak of about 360,000 in 1958, just before the longest drought on record (7.4 years, Figure 9.1; see also Box 9.1). In between there were another three severe droughts that collectively totalled 3.3 years.

Although there had been droughts in the past, the period spanning these five droughts and the dramatic rise in cattle numbers were probably the most critical ones for native mammals because they were so much more vulnerable to predation by dingoes and competition from exotic herbivores. Their difficulties were probably exacerbated by widespread, severe grassfires that became more frequent after Aboriginal peoples moved into settlements in the 1930s and 1940s, and were made even worse by the grassfires that burned most of central Australia in 1951 (Figure 9.1).

Dingo abundance and diet before and after the arrival of rabbits and cattle

Before the arrival of rabbits, the main prey of dingoes were probably medium-sized bandicoots and marsupials in flush periods and large kangaroos in droughts. Some of these prey are still eaten today even though they are considered rare, threatened or endangered. For example, bilbys were recorded in dingo stomachs and faeces in central Australia in the 1970s and 1990s, and spectacled hare-wallabies and nail-tailed wallabies were recorded in dingo stomachs in the Barkly Tableland in the 1970s.

Dingo numbers increased, perhaps trebled, with the pastoral industry, and did not decline in droughts as did other predators and prey because they were buffered against starvation by the increasing supply of cattle carrion. As already mentioned, cats and foxes could not compete against groups of dingoes for carcasses.

Although dingoes would have changed their diet to include more large kangaroos in drought, this food source would have been secondary to cattle carcasses. Red kangaroo populations did not substantially increase until the next decade or so when cattle grazing modified the growth pattern of native grasses to provide an increase in kangaroo food. This modified vegetation is commonly known as the 'marsupial lawn'.

How an increase in dingo predation could have led to the extinction of some native mammals

During drought the native fauna declined, and so did foxes, cats and rabbits. In contrast, dingo populations remained high thanks to cattle carrion and water provided for stock. However, dingoes would still have hunted live prey, particularly macropodids and bandicoots, their main prey at the time. Predation on them would have become increasingly severe as dingo populations grew and as their protective shelters were removed by cattle and rabbits.

The open plains, which once supported stands of long perennial grasses, would have been depleted by cattle and rabbit grazing, and stripped bare during drought. Any creatures depending on long grass or small shrubs for shelter were thereby rendered homeless and vulnerable to predation; they could not survive (Table 9.1). It seems, therefore, that the dingo was the main culprit in the extinction of many fauna in central Australia, particularly those that were medium-sized.

Feral cats and foxes as exterminators of native fauna

In the past two decades or so, cats and foxes have also been implicated in the extinction of small isolated populations of native mammals. For example, a single fox eliminated a small remnant population of the rufus hare-wallaby in the Tanami Desert in 1987, and feral cats killed about half the 56 hare-wallabies that had been reared in captivity and then released there. But, unlike the predation pattern of 40–50 years earlier, this kind of predation is possible only because rabbit populations are now high and allow high populations of foxes and cats to persist.

Feral cats are considered by many to be very efficient hunters and it is often claimed that cats concentrate on a small remnant of species and eventually kill all of them. This happens only if the cats have a backup source of food such as the rabbits in the Tanami Desert or as inadvertent 'handouts' supplied by humans in south-west Victoria, where the eastern barred bandicoot is reputedly under grave threat of extinction by cats. Without such staple food sources cats are likely to die before the prey, as indicated by the following example.

In Scotland's outer Hebrides feral cats catch rabbit kittens more successfully than they catch adult rabbits. However, rabbits do not breed in winter, and only the feral cats whose territories span the best hunting grounds can survive the winter. All other cats in adjacent areas die, despite the presence of thousands of overwintering barnacle geese, ground-nesting fulmars, and carrion that washes up along the beaches.

It is, therefore, interesting to compare the ability of cats to survive the Hebridean winters with that of cats in Australian droughts. Both populations have difficulty in surviving, but the Hebridean cats probably do better because they do not have to contend with dingoes.

In conclusion to this section, the question must be asked: If the Tanami Desert hare-wallabies and other remnant populations have survived the earlier ravages by cats and foxes, why not now? The answer may well be to do with minimum viable populations but it may also involve the higher rabbit populations that now buffer cats and foxes from dingo competition.

INTERACTIONS IN FAIRLY PRISTINE ECOSYSTEMS

Tropical coastal wetlands

At Kapalga in Kakadu National Park, on the coastal wetlands of the Northern Territory, most dingoes belong to stable packs of 3–8 members and defend more or less fixed territories. The main prey of dingoes are dusky rats, magpie geese and agile wallabies, and whenever one of these prey is temporarily unavailable dingoes switch to a range of at least 33 species of substitute prey although they usually concentrate on only one or two species at a time.

Wallabies are available all year round whereas the supply of rats and geese varies with wet and dry seasons. Although geese breed in the wet months (Jan–Mar), most are eaten as fledglings in the dry months (May-Nov) when water levels are low and dingoes can reach the floodplains. Rats also breed in the wet season and many are eaten then, but when rats erupt into huge plagues and cover the floodplains about every 3 or 4 years, most are eaten during the dry months when dingoes can get at them.

There is in fact a negative and significant relationship between the dietary intake of floodplain fauna (rats and geese) and forest fauna (wallabies and possums): the former are mostly eaten during the dry months and the latter in the wet months (Figure 9.2a).

Climatic conditions, therefore, influence both when and where dingoes hunt particular prey species. This alternation of predation between habitats, illustrated in Figure 9.2b, is a well-defined, predictable cycle in which dingoes do not appear to influence the abundance and diversity of any particular prey. In effect, dingo predation has achieved a balance with nature.

In other coastal regions of the northern wet-dry tropics where geese are present but do not breed, the model's basic components probably are consistently available prey (agile wallabies) and opportunistic prey (dusky rats); see Box 9.2. Further inland in areas where magpie geese and dusky rats are absent, the basic components probably are agile wallabies with possums and bandicoots in wilderness areas, and macropodids with cattle in pastoral areas.

Temperate coastal mountains

In south-east Australia dingo-prey interactions are determined more by wildfire than by rainfall. Most fires are low to moderate intensity so that the environment remains fairly stable and food supplies for dingoes are usually high. Fires of

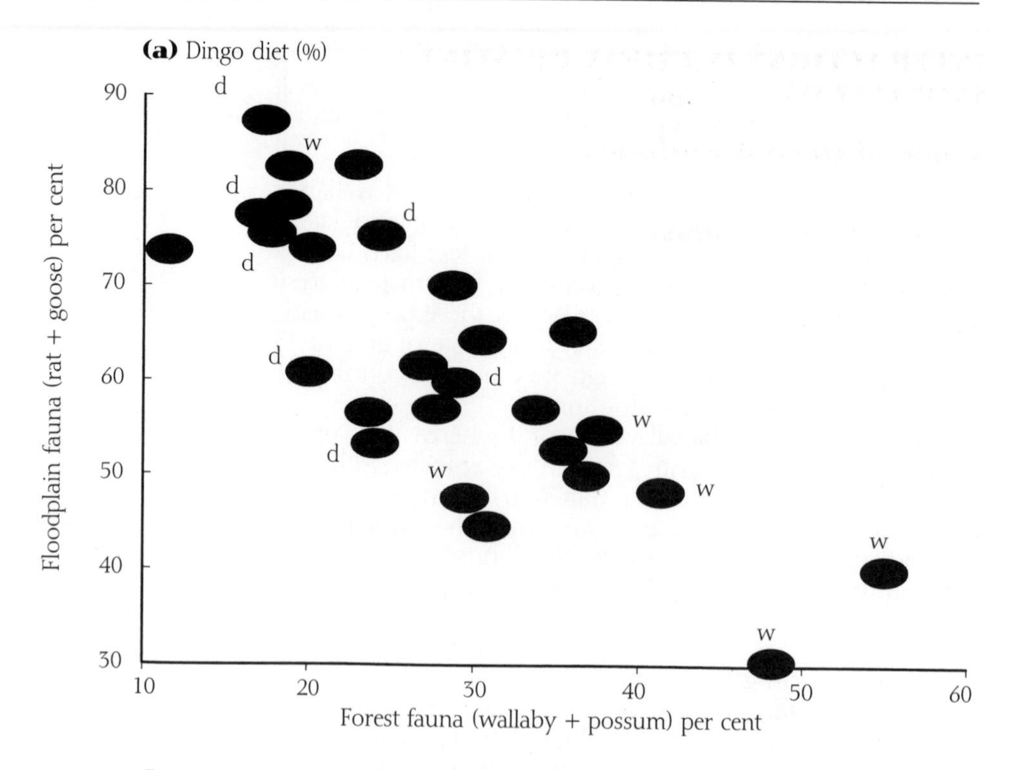

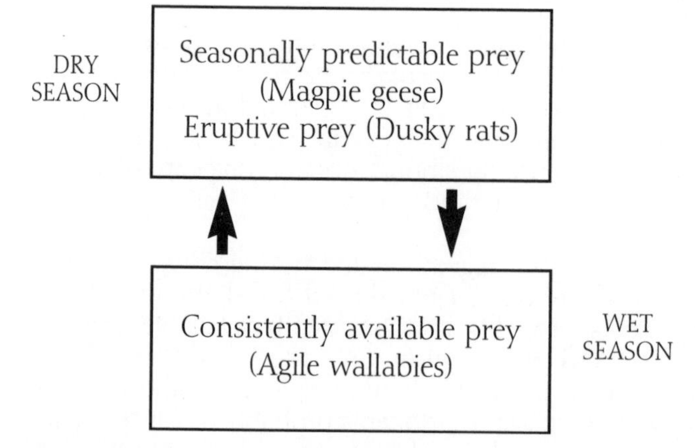

Figure 9.2

How dingo predation alternates between habitats in tropical Australia. (a) Switching from forest to floodplain habitats according to annual wet (w) and dry (d) seasons is triggered by rainfall. Dingoes cannot gain access to the plains until they dry out.
(b) A general model of dingo-prey interactions: dingoes switch between seasonal prey (rats and geese) and consistently available prey (agile wallabies) depending on which are available. When their main prey are temporarily unavailable dingoes substitute one or two species from a range of at least 33.

Box 9.2
Prey types defined

Main prey (primary prey): the most commonly consumed species.

Consistently available prey (staple prey): the most consistently though not necessarily the most commonly consumed prey; i.e., the prey dingoes can rely on.

Substitute prey (secondary prey, alternative prey): an alternative whenever main prey are unavailable.

Supplementary prey (ancillary prey): generally eaten to augment the staple diet; they fluctuate from being a major diet item to a minor one.

Opportunistic prey: prey that provide predators with a 'feast or famine' and thus cannot be relied on; these usually consist of small mammals that are either rare or in plague numbers.

Seasonally predictable prey: these are particularly vulnerable to predation at the same time and place every year, e.g. colonial ground-nesting birds; may be main prey or supplementary.

Scavenged prey: dead animals, the source of which is reliable in certain circumstances (e.g. cattle carcasses in droughts) but not in others (e.g. seabirds washed up on beaches); can be considered as a special category of opportunistic prey.

high intensity, although infrequent, devastate entire forests and change the prey base (variety of food) available to dingoes, but the total food supply usually remains high. It is, therefore, not surprising that the predatory cycle is less defined than in the tropical wetlands. Lower densities of dingoes living in smaller packs (averaging 3 members), and in smaller territories, are also a consequence of the environmental stability.

However, this predatory cycle still alternates between consistently available prey and seasonal prey. At Nadgee Nature Reserve, for example, the main prey are the medium-sized mammals (wallabies, rabbit, possum), and waterbirds (swans, coots) are eaten seasonally. Both prey types are supplemented by large macropodids (grey kangaroo) and small mammals.

In this ecosystem dingo predation sometimes affects prey diversity, abundance and population structure, probably because severe wildfires change habitats and thus alter the composition of the prey base. This encourages dingoes to either concentrate on a vulnerable, relatively uncommon species, or to prey on an abundant species, thereby relieving the pressure on another species and allowing its population to recover. At Nadgee, dingoes concentrate more than usually on macropodids immediately after severe fires and sometimes eliminate local pockets of grey kangaroos.

However, the effects of such intense predation is alleviated when waterbirds are in great abundance, as they are sometimes when severe storms replenish the coastal lakes with water and food for the birds. When there are no substitute prey available after severe fire, dingo numbers decline soon after the macropodids decline, in a negative feedback reaction.

Higher in the mountains where no waterbirds are available — at Kosciusko National Park for example — there seems to be no well-defined cycle. Dingoes hunt mainly wombats, wallabies and rabbits, and this diet is supplemented by a variety of other species. Similarly, in the mountains near Armidale in north-east New South Wales, the main prey of dingoes are macropodids, especially swamp wallabies, red-necked wallabies and eastern grey kangaroos. Dingoes concentrate on juveniles (pouch young and young at foot) and reduce their recruitment rates so considerably that the populations of these species decline; in some areas small isolated populations of grey kangaroos and red-necked wallabies were completely eliminated. This only happened, of course, because enough substitute prey were available to support the dingo population, otherwise the dingoes would have moved away or starved.

Another outcome of the dingoes' concentration on swamp wallabies was a disruption of the usual seasonal pattern of wallaby births. Many females ejected their pouch young when pursued by dingoes, but as most of these offspring were soon replaced (96% of sexually mature females carried a blastocyst), there was a continuous output of young instead of the usual spring-summer peak. Besides this change in breeding pattern, the number of ovulations per female changed, and so did male swamp wallabies' testicular and epididymi (mass of ducts leading from the testes) weights.

INTERACTIONS IN DISTURBED ECOSYSTEMS

How have ecosystems been disturbed? Two major factors responsible for this are the introduction of exotic animals, especially rabbits and stock, and pastoral industry infrastructures such as artesian bores and dams. The consequent alteration in habitats has made some native prey species increase but others decrease. Dingoes have mainly benefited from extra supplies food and water, which have helped them to survive drought and increase their numbers, but these extra

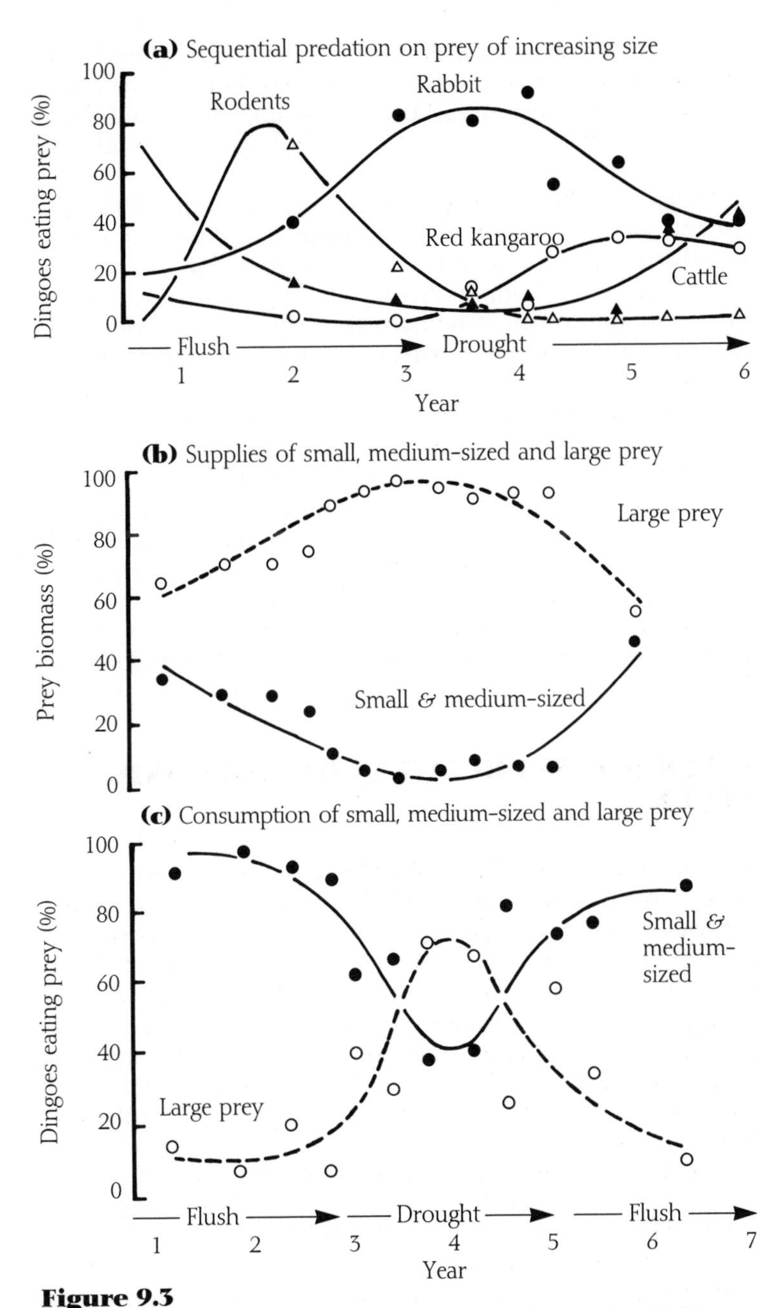

Figure 9.3
Dingo predation in disturbed ecosystems in arid Australia. (a) Dingoes eat prey of increasing body size as aridity increases, starting with small rodents, switching to rabbits and kangaroos as drought deepens, then relying on the carcasses of cattle that starve in a long drought. Dingoes focus on the most plentiful prey which, in terms of body size, fluctuates as the climate does, so that (b) supplies and (c) consumption of small and medium-sized prey is greater than that of large prey in flush periods, but vice versa in drought.

resources have also changed the natural pattern of predation. The interplay between seasons (drought and flush years), native prey and introduced prey (pests and stock) and dingo predation is well illustrated by a study made at Erldunda in central Australia.

When rains broke the longest drought on record (1958–65), rodents erupted over widespread areas and dingoes concentrated on them for about a year (Figure 9.3a). The rabbits came next and predominated in the dingo's diet for 3 years. When another drought reduced the rabbit populations, predation on red kangaroos increased, even though they became uncommon. Then, as this drought lengthened, cattle began to die and carrion became more frequent in the diet.

This sequential emphasis upon vertebrate prey of increasing body size as aridity increased is summarised as a general model in Figure 9.3 b–c, which indicates that dingo predation is greatest on small and medium-sized mammals during flush periods and greatest on large mammals in drought. This predation variance raises the question of what impact dingoes may have on prey populations. Are there particular critical periods when dingoes kill cattle? Does predation regulate or limit prey populations? These questions are now examined.

INTERACTIONS BETWEEN DINGOES AND STOCK

Dingoes do kill and eat cattle, sheep and other stock; they always have and always will. Pastoralists have so feared dingoes that many millions of dollars have been spent over the past 150 years or so trying to kill them or exclude them from pastoral areas. The longest fence in the world is a monument to that. Probably more has been written about the dingo wars than any other aspect of 'dingology', yet only two facts stand out: all the effort has been extremely expensive and, by and large, it has not worked. Are there alternative solutions?

For sheep farmers, probably not. Keeping guard dogs (e.g. the Pastore Abruzzes Maremma, Castro Laboreiro, Akbash, Anatolian Shepherd Dog) in the paddocks with sheep is potentially a low-cost option that may deter dingoes on small holdings. These dogs are reputed to deter coyotes in America and wolves in Europe. So far, the method has not been adequately tested in Australia but it is unlikely to work well in the large sheep holdings that run into thousands of square kilometres.

Fencing off dingoes from sheep seems the most practical solution, especially if areas adjacent to sheep paddocks retain adequate native species to support dingo populations, or if appropriate buffer zones surrounding sheep paddocks are maintained, especially those abutting National Parks and other wilderness areas.

Studies of dingo movements indicate that buffer zones should be 10–20 km wide (i.e. about two dingo territories wide) and that resident packs should be removed. Although this creates a dispersal sink (Chapter 8), it allows those dingoes moving in from further out to settle, and be exposed to control techniques *before* they reach sheep. Logistically, the most effective strategy is to (poison) bait the buffer zone periodically to reduce dingo numbers and thus maintain a dispersal sink; this has the added benefit of alleviating the need for frequent monitoring of dingo population levels. This strategy, based on Thomson's research in the Fortescue River region, has proved successful after its general adoption in Western Australia.

For cattle farmers, however, there are alternatives, as follows.

UNDERSTANDING DINGO PREDATION TO MAXIMISE CATTLE PRODUCTION

Scientists and cattlemen usually know what dingoes eat but this knowledge alone is not enough to understand why, how and when dingoes kill stock. Diet is usually determined from scats or stomach contents, both of which merely indicate the dingo's last feed. Also, dingoes sometimes kill stock but do not eat them, so dietary information can be misleading. Pastoralists cannot constantly monitor dingo activities around cattle herds for signs of trouble, so how can they predict when dingoes are likely to be harmful or useful, and how can they decrease the competition from wallabies, pigs, rodents and rabbits to minimise stock losses?

The important sets of information are:

1 The composition of the prey base on which dingoes survive, how it varies between regions (e.g. coastal wet-dry tropics, inland wet-dry tropics), and how it varies within regions depending on climate (wet and dry seasons, drought and flush runs of years).

2 The social status of the local dingo population — this greatly determines the nature and extent of predation.

Various aspects of these information sets are affected by cattle management practices, dingo control techniques and other factors such as the proportion of dingo/domestic dog hybrids in the dingo population. The following two cases demonstrate this.

Case 1

In the Barkly Tableland of the Northern Territory, the main items of dingo diet are small mammals (mostly long-haired rats) and cattle, supplemented by an assortment of lizards, birds, insects and macropodids. In years when there are plagues of long-haired rats (about 1 in 9 years) dingoes eat proportionally more rats as rat numbers increase (functional response) and more dingoes survive (numerical response). In the years that immediately follow rat plagues, dingoes keep their numbers up by eating cattle but eventually they decline, mainly from starvation but sometimes from epizootics of canine hepatitis and distemper.

Although dingoes sometimes form groups, similar to packs, such groups are congregations of mostly unrelated dingoes sharing resources — water and sometimes cattle carcasses. Away from these resources, most dingoes operate independently. However, in the dingo mating season (which lasts about 4 months, peaking in March/April), dingoes form temporary breeding groups which often comprise one oestrous female and several randy males. Within these groups there are many aggressive interactions but actual fighting is uncommon because of complex behaviours associated with dingo dominance hierarchies. Aggressive behaviour can be appeased or diverted by submissive behaviour to avert serious wounding and death.

In the Barkly Tableland, the dingo mating season coincides with the peak in calving and this coincidence contributes to the deaths of many calves. Calves and dingoes (usually in breeding groups) are often together at water. Calves become frisky and dingoes pursue and attack them, perhaps because of their high levels of breeding hormones at this time. In many cases, such attacks are probably more of a displacement activity than a hunger drive, perhaps because dingoes become frustrated from competing over oestrous females and fighting with rival males.

However, a calf cannot appease or divert the aggression as a submissive dingo would, so the dingo, irrespective of social rank, continues to attack, often joined by other dingoes, until the calf becomes immobile or dies. Calves killed this way are rarely eaten (see Plate 20). Even if the calf survives well enough to be sent to market, the meat is often classified as second class because of scars from dingo bites. Such loss to the cattle industry is probably substantial but it is difficult to estimate; because calves are seldom eaten by dingoes, examining dingo stomachs or faeces would be misleading.

On the Barkly Tableland, therefore, a more appropriate management option would be to *manipulate cattle herds so that fewer calves are born in the dingo mating season*. Such an option is apparently feasible.

Case 2

In the Alice Springs region of the Northern Territory, calf losses such as those just described are rarely recorded, probably because the peak of the calf drop does not coincide with the dingo mating season here — most calves are born in September and October. Although droughts in this region are unpredictable in frequency and severity, the effects of dingo predation are predictable and pastoralists can benefit from this factor in several ways, as follows.

During runs of flush years, dingoes eat mostly small and medium-sized prey, mostly rabbits and rodents (Figure 9.4). Dingoes maintain high numbers then because human control methods effectively increase the numbers of breeding dingoes and because abundant natural food ensures the survival of most pups. During the middle and latter years of drought, dingoes survive very well on cattle, usually carcasses. Most calves killed by dingoes during these periods do not incur economic losses because calves cannot survive long droughts. In fact, such calf losses are probably an advantage to pastoralists (Figure 9.4) because the mother's chance of surviving the drought is increased if she has no calf to feed. Thus dingoes benefit pastoralists by helping breeders survive droughts and the management technique is easy: *relax dingo control in droughts.*

Management decisions during the initial periods of droughts are tricky, however. After a few dry months the populations of small native mammals and rabbits decline before cattle begin to die. Dingoes therefore have no choice but to tackle large kangaroos or kill calves, so dingo control is necessary. If the drought breaks after a few months, such calf losses are indeed economic losses. Since the losses are beneficial if the drought continues, when should a pastoralist heighten and relax dingo control? One pastoralist in central Australia is reported to have shot calves, not dingoes, after the first 7 months of a drought, and also to swap truckloads of his calves for hay from farmers in southern Australia. It was also reported that this property recovered from the drought much earlier than neighbouring properties did, which suggests that dingo predation and destocking both work well.

Dingoes may also assist pastoralists immediately after drought. Rabbits and small native mammals quickly resurge when the rains come, and dingoes concentrate on them again. If dingo numbers are high due to relaxed dingo control in the drought, the heavy predation will hold back the rabbit numbers so that they take longer (perhaps 2 years) to reach their former levels (Figure 9.5). This gives a pastoralist more time to establish other forms of rabbit control (e.g. warren ripping and releasing myxoma virus) before they become a serious problem again. In effect, dingoes and pastoralists work in tandem this way to control rabbits and other grass-eating competitors of cattle.

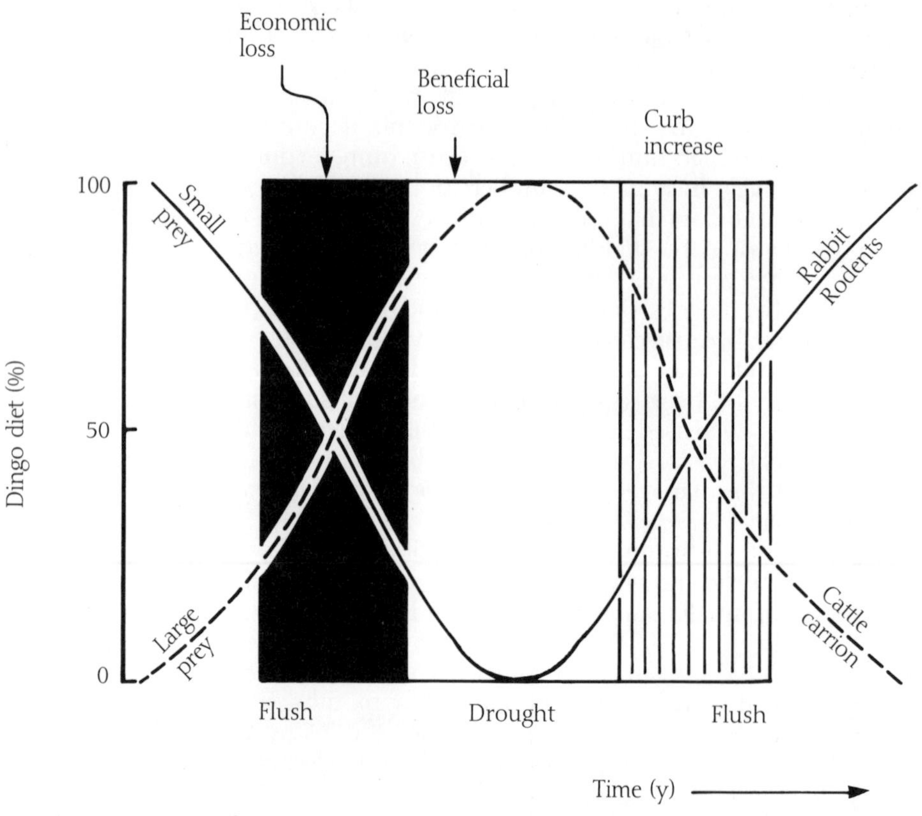

Figure 9.4

The good and bad of dingo predation in arid Australia. Understanding and manipulating dingo predation on stock and native species (incl. rabbits) helps pastoralists to minimise calf losses, improve cattle breeding and increase grass supplies. In flush periods dingoes mostly eat rabbits and rodents, the major grass competitors, so it is not the time to kill dingoes. In the first months of a drought when rabbits and small native prey decline, dingoes kill calves which might have survived a short drought (economic loss), so troublesome dingoes should be killed then. If drought continues, suckling may reduce their mothers' chances of survival, so dingoes should be allowed to kill calves then (beneficial loss). Straight after the drought rodents and rabbits multiply and dingoes switch back to them. Dingoes sometimes curb these increases for up to 2 years which gives pastoralists time to set up other controls, e.g. myxomatosis or warren ripping.

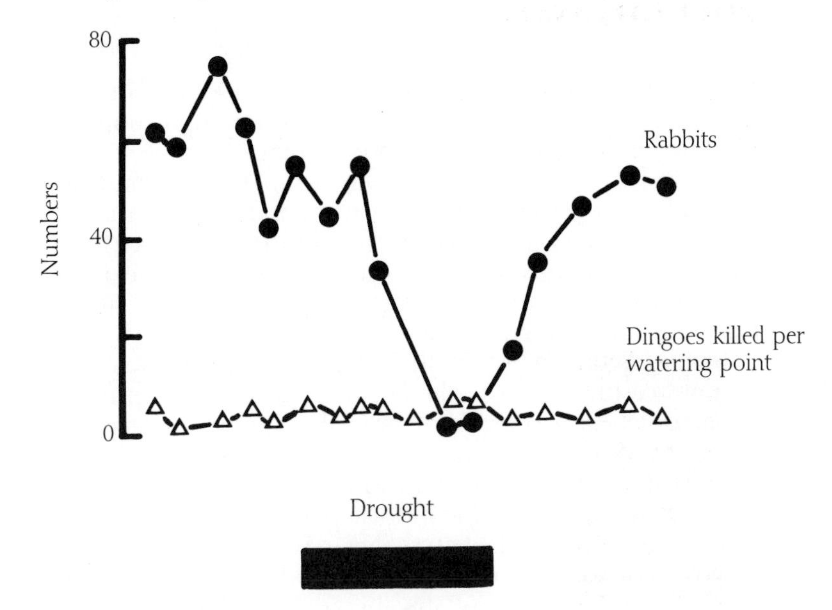

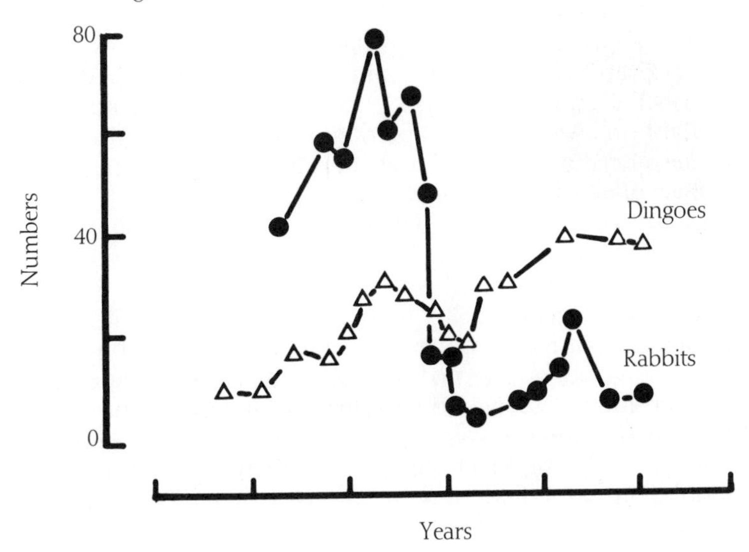

Figure 9.5
Comparison of two dingo populations in arid central Australia: (a) one was subject to intense trapping, shooting and poisoning; (b) the other was not. After a severe drought rabbits resurged in the habitat of population (a), but in the habitat of population (b) they took about 2 years to regain their pre-drought levels; dingo predation had apparently curbed rabbit numbers there.

DOES DINGO PREDATION REGULATE PREY POPULATIONS?

Dingo predation has often been said to be the cause of fluctuation or lack of fluctuation in some prey populations, as indicated by the following seven examples.

Control by dingo predation was assumed to account for the contrast in density of red kangaroos and emus on the two sides of the dingo fence between the borders of Queensland, New South Wales and South Australia. Outside the fence, where dingoes abound, kangaroos and emus are rare; but vice versa applies inside the fence.

In the mountains of north-eastern New South Wales the occurrence of swamp wallabies in dingo diet was proportionally higher than expected from the number of observed wallabies. When dingo numbers increased, so too did the consumption of wallabies (52% to 73%) which was soon followed by a marked decline in their population.

At Nadgee in south-eastern New South Wales, post-fire dingo predation on macropodids held their numbers in check for 2–3 years; probably because the fire opened up habitats and made prey more vulnerable.

In arid central Australia red kangaroos became more vulnerable to predation during drought partly because they were clumped around waterholes and the remaining feed. At one site on the plains, kangaroo populations declined during a drought in which dingo predation became progressively greater, and they remained low after drought — at about 15% of their pre-drought level.

In the Harts Ranges red kangaroos declined from being common to very rare after a 7.5-year drought, and populations did not recover in the subsequent 10 years even though pasture was generally better than average. Dingoes, which had become increasingly common as rabbits became abundant after the drought, seemed to have mediated competition between rabbits, cattle and kangaroos to the detriment of the native herbivore.

In the Fortescue River region of north-west Australia euro populations were fairly low in an area where dingo populations were allowed to remain high, because they preyed selectively on particular age classes of euros. When dingoes were greatly culled by a poison baiting program, euro populations immediately and dramatically increased.

In Queensland analyses of bounties paid annually on dingoes and pigs over 24 years indicated that pig mortalities increased more than threefold with every doubling of dingo numbers, and this inverse relationship suggested that the numbers of dingoes were directly responsible for the pig numbers.

Box 9.3

Regulation, limitation and control: definition and assessment

Regulation is the process whereby a population returns to a stable equilibrium in which loss (death and dispersal) equals production (births and immigrants). Prey populations are regulated by predators when predation has a direct density-dependent effect, usually over a range of low prey densities. Vertebrate predators can also affect small mammal populations in a delayed version of this process.

Limitation is the process that sets the equilibrium point; its density-dependent and density-independent factors determine the population's gains and losses. Predation can check prey populations.

Control is the process whereby human techniques keep pest populations at an arbitrary level that is acceptable to pastoralists and conservationists.

Regulation can only be demonstrated by experiments that alter predator or prey densities and test for density-dependent effects. The required evidence is that disturbed (increased or decreased) populations have returned to the unaltered level of the control population. Most experiments so far have involved predator removal, where regulation can only be verified if (a) the prey population increases after predators are removed but (b) does not return to its original density after predators return to their previous level.

Alternatively, in experiments that increase prey density, regulation is demonstrated if prey mortality, as a percentage of the prey population (the total response), can be positively correlated with prey density.

These examples show that dingo predation certainly affects prey populations, especially macropodids, but its long-term effect and whether or not the effect is regulatory cannot be fully assessed. Its regulatory effect (or the absence of any) can only be demonstrated by experimental studies (Box 9.3) but, to date, only one relevant study has been conducted on dingoes.

Pig predation by dingoes in the tropics — regulation or limitation?

An experiment was conducted at Kapalga in northern Australia to examine whether or not dingo predation regulated populations of feral pigs. As explained in Chapter 9, the main prey of dingoes at Kapalga are dusky rats, magpie geese and agile wallabies, with a large range of substitute prey

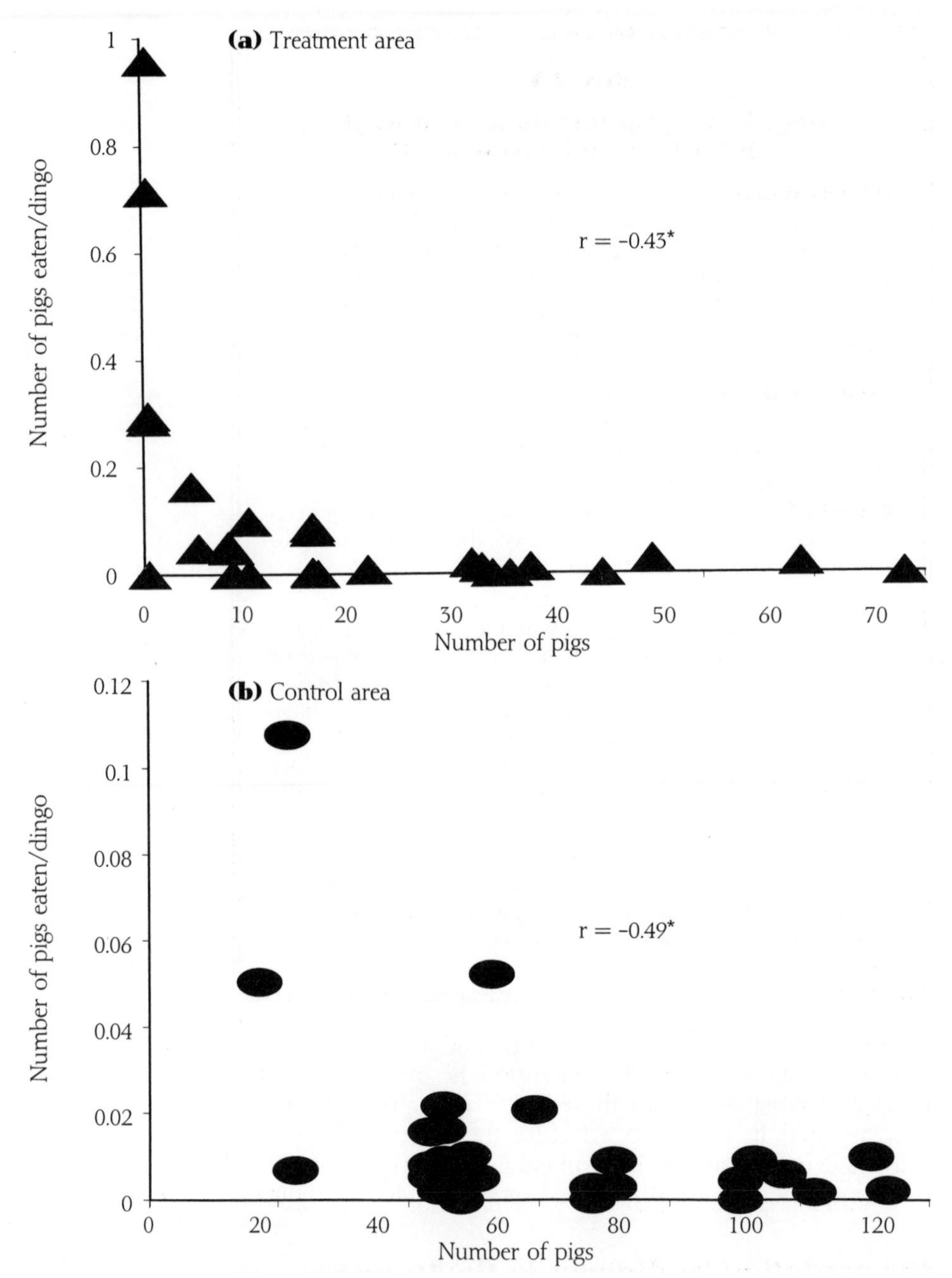

Figure 9.6

The rate of predation on feral pigs: (a) pig numbers increased due to an experiment; (b) they remained fairly constant (from Corbett, 1994). Since the number of dingoes remained fairly constant in both areas, the predation rate (total response) corresponded with the functional response (no. of pigs eaten per dingo). Both relationships were depensatory (inversely density dependent) so that dingoes did not regulate these feral pig populations.

including pigs. After feral swamp buffalo had been removed from half the 614 km^2 study area, the number of pigs doubled and there was a threefold increase of pig in dingo diet. Dingo numbers remained fairly constant throughout the study.

The rate of dingo predation on pigs was then examined in areas with and without buffalo (control and treatment areas respectively) to test two hypotheses: (1) dingo predation alone regulates the pig population; (2) the pig population is limited by interference-competition from high densities of buffalo and other factors including depensatory (inversely density-dependent) dingo predation. The critical prediction for hypothesis (1) was that dingo predation, as a proportion of pig numbers (i.e. the functional response), would increase; whereas hypothesis (2) predicted that such predation would decrease. Data collected over 7 years indicated that the dingo predation rate (functional response) was negatively correlated with pig abundance for both the treatment and control areas (Figure 9.6). This clearly indicated that dingoes were not regulating the pig population. Instead, dingo predation probably acted in concert with the interference-competition from buffaloes. This form of competition occurred during the dry season when buffaloes and pigs congregated around the dwindling waterholes. Stamping by buffaloes hardened the ground and this greatly hampered the pigs' access to the subterranean food which was critical for their survival. This conclusion is supported by a finding that high numbers of

Box 9.4
Environmentally modulated predation and the regulation of prey populations

In semi-arid Australia droughts reduce many animal populations to low densities, and during these periods carnivores can regulate small and medium-sized mammal prey populations. Such environmentally modulated predation has been recently demonstrated with foxes and feral cats preying on the introduced rabbit, and inferred from dingo predation of rabbits. The theory is that rabbits are trapped in a 'predator pit' where their reproductive rate is too low for the population growth rate to exceed the predation rate.

During the period of lowest rabbit density, dingoes kill more large prey as substitutes for rabbits (e.g. red kangaroos and emus) but whether or not the large prey are regulated by dingo predation has not been demonstrated. Similar inferences have been drawn for feral pigs in tropical northern Australia where rabbits are absent, but the only experimental study to date clearly showed that dingoes did not and probably cannot regulate pig populations.

buffaloes are associated with low numbers of pigs.

Are there circumstances in which dingo predation alone can significantly curb pig and other prey populations? The evidence of studies made in other regions of Australia indicates that this can only happen if prey numbers are initially depressed by a widespread environmental event such as drought in arid rangelands or intense wildfire in temperate forests. At the same time, the extra food that such circumstances usually provide allows the predators to survive them. For example, in arid central Australia cattle carrion enables dingoes to survive droughts. The scene is thus set for predators to concentrate on a substitute prey and perhaps regulate their population. In this case dingoes eat red kangaroos as a dietary substitute for rabbits, thus regulating the kangaroo numbers. An equivalent situation would have to exist before dingoes could regulate pig populations (Box 9.4).

According to the environmental switching model for dingo predation at Kapalga (see Figure 9.2b), populations of all the dingoes' main prey (rats, geese, wallabies) would have to decline simultaneously so that dingoes would be obliged to eat pigs. This is extremely unlikely (it would probably occur once in 175 years) because rat populations decline to low levels about every 3.5 years, young magpie geese are unavailable about every 5th year, and agile wallaby populations decline perhaps every 10 years in response to a particular fire cycle.

Even if such events did coincide, the dingo population would most likely emigrate or starve because of the lack of substitute food such as cattle carrion. Therefore, in tropical regions of Australia, with the absence of prolonged droughts or other circumstances to simultaneously reduce populations of main prey, *it seems unlikely that dingoes can ever limit feral pig populations to levels that are low enough to satisfy pastoralists and conservationists.*

CHAPTER 10

THE FUTURE OF EXPATRIATE DINGOES

The greatest threat to the survival of pure dingoes throughout the world is crossbreeding with domestic dogs, which swamps the pure dingo gene pool. This chapter examines the nature and seriousness of this threat and whether anything can be done to reduce it.

WHERE ARE THE WORLD'S LAST BASTIONS OF PURE DINGOES?

The world's remaining dingo populations are concentrated in south-east Asia and Australia. Surveys conducted over the past 20 years in those areas indicate that Thailand presently has the purest populations of dingoes in the world (Table 10.1), although the situation appears to be rapidly changing. No hybrids were recorded in Thailand when the first surveys were conducted in 1975, whereas 49 hybrids were recorded in a sample of 632 canids taken 9 years later. Even allowing for flaws in sampling procedures, this 8% increase indicates that hybridisation is on the increase and probably has something to do with Thailand's rapid 'modernisation' over the past 40 years or so.

Many Thais are now adopting many aspects of Western culture, including the ownership of domestic dogs as status symbols. There is now a National Canine Association in Thailand to accommodate the rapidly rising interest and demand for imported domestic breeds as well as the distinctive Thai ridgeback (Box 10.1), and international dog judges regularly visit Bangkok. Unfortunately, the upshot is that many of these domestic dogs will inevitably 'escape' from society and breed with the pure dingoes. Thus, processes similar to those described in the next section concerning dingoes in Australia will be set in motion, and eventually lead to the extinction of Thai dingoes.

Table 10.1

The approximate proportion of pure dingoes remaining in various countries.

Thailand and Australia are the last bastions. There is a significant and positive relationship between the number of years a country has been colonised or received Western influence (signifying domestic dogs) and the proportion of hybrids in that country ($r = 0.54$, $n = 8$, $P<0.01$)

[a] No. of years domestic dogs have been assumed present
[b] Classification based on a combination of coat colour and other physical traits
[c] Pure dingoes assumed present, from other evidence (e.g. photographs)

Country	Years since first colonial settlement or Western influence[a]	No. canids examined	Dingoes %
Thailand	c. 40	118 skulls	98
		632 external[b]	92
		1078 pelages	87
Australia	206	1591 skulls	87
		1535 pelages	86
Myanmar	168	101 external	50
Malaysia	c. 200	13 external	42
Indonesia	399	398 external	26
Laos	95	38 external	13
Philippines	429	237 external	5
India	229	10 external	1
Borneo	c. 200	0	present[c]
New Guinea	111	0	present
Cambodia	131	0	present
Vietnam	192	0	present

DINGO–DOG HYBRIDS IN AUSTRALIA: IS EXTINCTION IMMINENT FOR PURE DINGOES?

After the European settlement of Australia, dingo numbers rapidly increased as a consequence of an increase in food and water supplies, and because human control measures allowed more females to breed. The total dingo population in Australia probably peaked during the 1930s, 40s and 50s, and since then numbers have remained high but the proportion of pure dingoes in the overall population has declined at an alarming rate.

Dingo samples collected in the 1960s and 1970s indicated that about half the wild canid populations in the more closely settled regions of southern Australia were hybrids, although populations in northern Australia and other remote areas were essentially pure dingoes (Figure 10.1).

That was several decades ago. The situation is almost certainly worse now and hybrids probably occur throughout Australia, even in the most remote areas. The most recent survey (1981–85), made in north-east Victoria, confirmed the trend of increasing hybridisation there, and if this continues at the present rate, pure dingoes may well be extinct before the end of the 21st century (see Figure 8.2).

How did this situation eventuate and why has the problem become worse in recent years? The reasons are readily apparent in the more settled regions of southern and eastern Australia, but less so in other, more remote regions.

Hybridisation in settled regions

Southern and eastern Australia have become even more settled in recent years, with an associated increase in industry (forestry, mining), tourism and other recreational activities

Box 10.1
The Thai ridgeback dog

In central and east-central Thailand there is a dog known as the 'Siamese ridgeback' or 'saddleback dog'. Most resemble dingoes and coat colours range from ginger, black, white, creamy, and blue-grey; some have a black muzzle. This 'breed', first described about 1910, came from Phuquoc Island (off Vietnam) and was possibly derived from a domestic dog, the Rhodesian ridgeback, earlier introduced into Indochina by the French and subsequently taken to Thailand. Other reports suggest that Phoenician sailors brought dogs to Indochina from southern Africa.

In the early 1900s the French used Siamese ridgebacks for hunting deer and other game, as did the local people. These dogs are still uncommon and command high prices ($150–300) from outlets such as Chatuchak weekend market in Bangkok. In Thailand, they are usually in demand as watchdogs, but they are now becoming fashionable status symbols. Siamese ridgebacks have also been exported to America where they are known as an unofficial dog breed called the 'Phuquoc greyhound'.

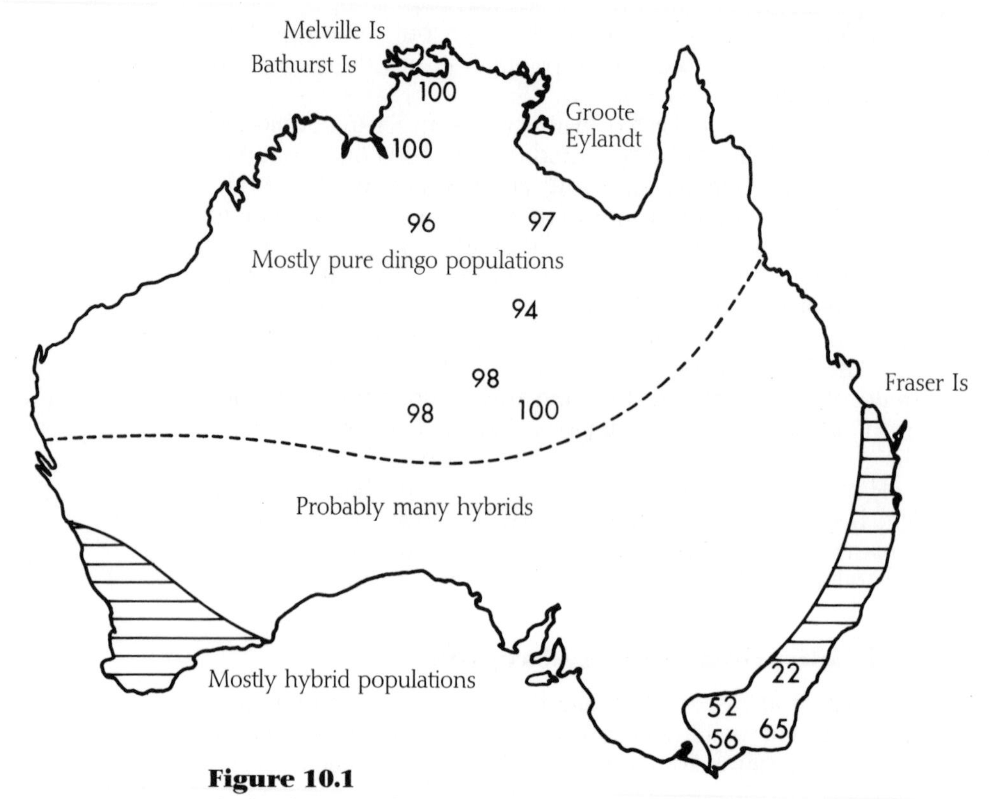

Figure 10.1

The last bastions of pure dingoes in Australia, according to the measurements of 1,611 skulls. The numbers refer to the percentage of samples classified as pure dingoes — the higher the number, the purer the population. Most data (from surveys made over 30 years ago) indicate that most pure dingoes (94–100% of samples) lived in the northern tropical regions, and all hybrids there were sampled near Aboriginal settlements or cattle station homesteads. There were no populations of pure dingoes in southern Australia (only 65% maximum). Although no samples have been collected in the intervening semi-arid regions, most dingoes there are probably intermediate, i.e., between the values for N and S Aust. Figure 10.2 indicates why there are probably fewer pure dingoes throughout all regions of Australia today. The most recent survey (Jones, 1990), confirms that hybridisation is increasing. There are few data from SW Aust., but the composition of wild animals there is probably similar to that in SE Aust.

(skiing, fishing, bushwalking), all of which demand easy year-round road access to the dingo's refugia in mountainous habitats. For example, in 1966 when I first began studying dingoes in north-east Victoria, the few roads above the snow-line were mostly logging tracks that required 4-wheel drive vehicles, and many of the doggers ran trap lines from horseback. Today you can drive on a bitumen road over many of the highest peaks in conventional vehicles, stop at wayside stops for a 'cuppa' and let the dog out for a wander.

Inevitably, the size and isolation of dingo refugia have continued to shrink and contact between dingoes and domestic dogs has increased to the extent that dingoes simply cannot avoid dogs any more. By nature, a dingo seeks its own kind to mate with, but without a pure dingo it will readily substitute an in-season domestic dog.

According to the following data, there are certainly plenty of feral dogs to choose from. In the 1960s, in the bush areas of Gippsland and north-east Victoria, doggers trapped many domestic mongrels but also many distinct domestic breeds including alsatian, kelpie, border collie, highland collie, Australian cattle dog (red and blue varieties), smithfield heeler, samoyed, bull terrier, labrador, deerhound, greyhound, cocker spaniel, fox terrier, curly retriever, wire-haired terrier, and even the regal corgi.

Where did these dogs come from? Many were probably accidentally lost by hunters, bushwalkers, picnickers, and the like. Some may simply have 'gone bush'. The identity tag of an alsatian showed it had escaped from a town 90 kilometres away and 11 years earlier.

There are many anecdotes told by doggers and other local residents of Gippsland and north-east Victoria about dogs being set loose in the bush some 3–4 decades ago, and there is no reason to doubt them. For example, alsatians were supposedly released there from a breeding kennel in the Buchan District, and by Italian tobacco and hop farmers along the Buffalo River — especially when the registration fee was raised to £5 (the equivalent of $10), a considerable sum of money in those days. Construction workers on the Snowy Mountains Hydroelectric Project apparently released huskies, samoyeds and other pet dogs when they completed their contracts.

It is said that graziers with cattle runs in the snow country (Bogong High Plains) used to collect stray dogs from nearby towns (including those from the dog pounds) and tether them to trees during cattle musters. Their excited barking apparently helped with yarding the cattle, and the dogs were released afterwards.

The upshot of these and similar incidents throughout the highlands of south-east Australia is that the high proportions of hybrids in southern and eastern Australia must inevitably increase even further, and dingoes there are most certainly doomed to extinction.

Hybridisation in remote areas

In more remote areas of Australia, such as the Top End of the Northern Territory, a similar sequence of events is beginning to unfold. The network of roads is expanding and being upgraded as a result of the booming tourist and mining industries, and the increasing need to freight goods and livestock more efficiently.

The Alligator Rivers Region (in Kakadu National Park) is a classic example. This area is designated world heritage and is the epitome of Australian tropical wilderness and pristine dingo refuge. When I first travelled to this region in the early 1970s the outward journey took two days via an indirect route that was possible only in the dry season. Today there is year-round access on a bitumen road and one can 'see' the sights of Kakadu from an airconditioned bus or hire car and return to Darwin in only one day! There is also a township in the heart of Kakadu, Jabiru, which presently boasts some 900 people and over 500 registered domestic dogs besides an unknown but large number of domestic dogs kept by the local Aboriginal people.

Similar developments are occurring in other parks and recreational areas in the Northern Territory, and Aborigines now tend to be more sedentary in larger camps with larger groups of dogs. Undoubtedly, dogs are taken into the bush more frequently now and dingo-dog contact has greatly increased in recent years, so it seems the stage is set for hybridisation to skyrocket. Yet there have always been such opportunities for hybridisation and so far few hybrids have been recorded in remote areas — none, in fact, from Kakadu. Why not? What has changed to make the situation more urgent now?

The main processes of hybridisation

The two most common situations in which dingoes and domestic dogs make contact are when dogs go bush, and dingoes come to town. However, since interactions between dingoes and domestic dogs in the bush differ greatly from those in urban places (Figure 10.2), so too do the rates of hybridisation.

For more than 100 years cattlemen, fishermen, bushwalkers and holidaymakers have taken their dogs out bush and occasionally lost them, and Aborigines have kept domestic dogs for much the same period. However, the behavioural differences between dingoes and dogs seem great enough to make it difficult for dogs to infiltrate dingo society and breed, particularly in remote areas where there are more dingoes. There is now a trend for people to (illegally) acquire dingo pups as pets. The pups may be easily handled, but the adults are not good pets, simply because they are wild animals (Box 10.2). A 'pet' dingo is likely to use its owner's home as a base from which to roam and do as it pleases, or else it is abandoned when it becomes an adult.

The upshot of this 'pet' trend is that dingo-dog contact is

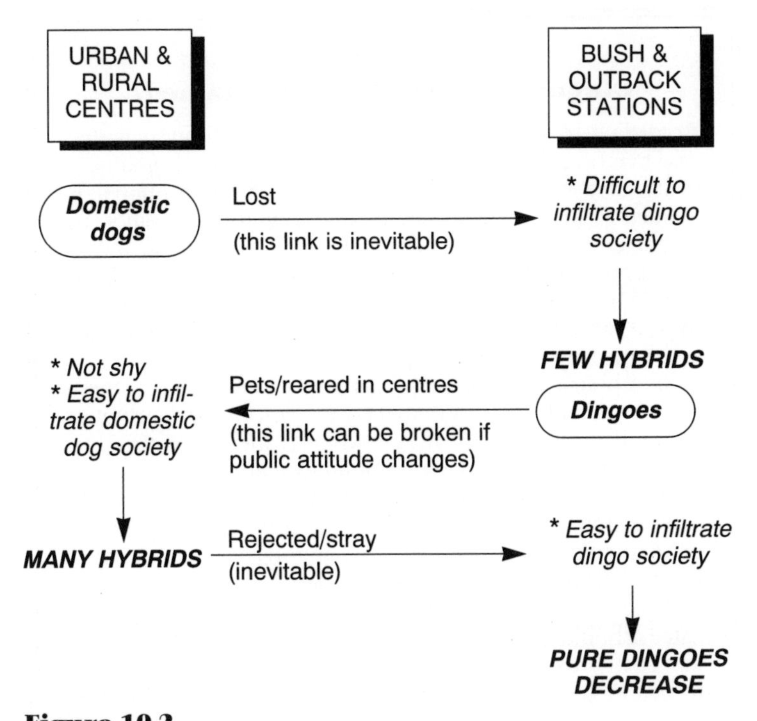

Figure 10.2
The main locations and processes of hybridisation between dingoes and domestic dogs.

increased; because pet dingoes have grown up in the urban situation without those social behaviours that curb breeding, crossbreeding is also markedly increased. Many such hybrids are rejected by owners or stray to the bush where their dingo genes make it easier for them to infiltrate wild dingo society and breed with pure dingoes. This process occurs more frequently in semi-rural areas outlying large urban centres — for instance, near Darwin (1991 population: 97,893) and Alice Springs (1991 population: 25,554). In these semi-rural areas 10% and 9% of the people, respectively, live on 2–5 hectare blocks where wild dingoes, 'pet' hybrids and free-ranging domestic dogs freely intermingle and often interbreed.

WHAT CAN BE DONE TO SAVE DINGOES?

What can be done to save dingoes? Should anything be done? Before attempting to solve any problem, it must first be acknowledged. Education is the best way to help people recognise the dingo as a native Australian (Box 10.3), understand its plight, and push for policies to retain the species as part of Australia's national heritage. But public education will

Box 10.2
Dingoes — domesticated or tamed?

According to the Oxford Dictionary, domestication is 'the selective breeding of (originally wild) species by humans in order to accommodate human needs'. A tame animal is defined as 'accustomed to man, or having the disposition or character of a domesticated animal'. Domestication, therefore, implies that the genetic makeup of an animal has been altered by human interference (selection) whereas the genetic makeup of a tame animal retains the original genetic makeup shaped by natural selection.

In a practical sense, the progeny of a long line of domesticated animals will 'automatically' perform the activities required by its human owner, whereas the tamed progeny of a wild animal requires retraining (retaming) with every generation. Thus, it is theoretically impossible to domesticate a wild animal and maintain its natural behaviour patterns, because two genetic makeups are involved and an animal has to follow one pattern or the other.

Wild dingoes, therefore, can be tamed but not domesticated. Should humans determine and selectively breed certain standards and characteristics for dingoes, they will cease to be dingoes. A domesticated 'dingo' is not a dingo but just another breed of dog.

achieve very little unless hybridisation is also dealt with by everybody — governments, concerned societies, and you.

Even if most people agreed that it is necessary to preserve dingoes, I wonder how many would also agree to give up their 'traditional right' to own a domestic dog or, if 'qualified' to have one, would agree to sterilise it? This would apply particularly to people living in wilderness areas and people on cattle stations, and to Aborigines living in the bush. And of course, the present state and territory laws that ban dingoes as pets for the general public would have to be maintained. These are all elements of the drastic remedy required, but everybody knows that they are unlikely to come about because laws banning people from doing something seldom work; they just get people's backs up.

Such has been the experience in America. People who keep wolf hybrids object vehemently to any move to outlaw these animals, few object to the idea of a permit, nor to the requirement that hybrids kept as pets must be neutered. Perhaps a similar philosophy could be adopted for keeping dingoes by interested members of the public. Domestic dogs, too, should be neutered in outback mining towns and Aboriginal settlements in wilderness areas such as Jabiru in Kakadu National Park.

Box 10.3
When is a native a native?

The Oxford Dictionary defines a native animal as one that is inborn, indigenous, or derived from one's country. In that sense the dingo is definitely not a native Australian, because, as explained in Chapter 1, dingoes evolved in Asia and were transported to Australia by Asian seafarers.

Yet during its time in Australia over the past 4,000 years or so, the dingo has interacted with indigenous animals, and responded to and changed aspects of the environment. For example, competition from dingoes probably caused the extinction of thylacines on the mainland, and the pressure of their predation has modified and influenced the subsequent evolution of certain ecosystems. Thus, in a functional and evolutionary sense, dingoes are as much a part of the Australian environment as, for example, modern kangaroos, which have evolved from the megafauna of yesteryear.

Defining a native Australian animal as one that arrived in Australia before European settlement (about 200 years ago) is also inappropriate because some animals that arrived in Australia after 1788 are now unique to Australia. For example, rabbits in Australia now physiologically and genetically differ from the original Spanish rabbits (first taken to England before coming to Australia), and the differences are due to their interaction with the peculiarities of the Australian environment. Since it is unlikely that Australia will ever be free of rabbits, does this mean that we have a native Australian rabbit? Similarly, feral cats and foxes have caused extinctions and modified environments, and will probably be in Australia forever; so, in a functional sense, can they also be considered as Australian mammals?

What about people? After World War II, immigrants from central Europe were labelled 'new Australians', but many of the very same people now 'patriotically' complain about the latest wave of Asian immigrants. Yet Asians (Macassans) were living in northern Australia long before the European settlement of Australia. Going back further in time, not even the Aborigines can be considered as the original native Australians by the dictionary definition because, despite the obscurity of their place of origin, anthropologists concur that the site was not Australia.

So, what constitutes a native animal? It might be as well to consider a native Australian as a species or subspecies that lives in Australia and has significant ecological and/or cultural impact, regardless of taxa, birthsite, race, language, length of time in Australia, and so on. On that basis, the dingo certainly qualifies as a native Australian.

State and territory governments can and do play their part here by recognising the overall role dingoes play in wilderness and pastoral areas, and by legislating accordingly. The Northern Territory removed dingoes from the pest list in 1976 but does not protect them, except in parks, reserves and in Arnhem Land. In South Australia, dingoes north of the dingo fence were declassified as pests (but not protected) in 1977, and remain proclaimed pests in the 40% of the state inside the dingo fence; bonuses on scalps, however, were entirely removed in 1990. New South Wales protects dingoes only in national parks and conservation estates, as does the Australian Capital Territory; and dingoes remain 'declared pests' subject to various levels of control throughout Western Australia, Queensland and Victoria. It is interesting that dingoes are also classified as vermin in Tasmania, even though they have never been found there. New laws on the status of dingoes and on the keeping of dingoes by the general public are presently being considered by most state and territory governments.

Governments can also use the growing database of dingo ecology to instigate rational, efficient control methods. The Northern Territory Government was the first to stop annual broadscale aerial baiting, and the Western Australian Government drew on new knowledge about dingo movements to set up buffer zones (two dingo territories wide) between pastoral and wilderness areas.

Dingo preservation societies

When a species becomes endangered it is usually preserved in a zoo or a wildlife park, but there are problems with this. For example, inbreeding and the 'unnatural surroundings' discourage the selection of wild characteristics. For the dingo there is an alternative: preservation societies. These small groups of dedicated people legally obtain dingoes to preserve dingoes and enjoy them. In essence they are equivalent to private zoos where dingoes are treated as companion animals. Two of the best known societies at the time of writing are the Australian Native Dog Conservation Society Ltd at Bargo in New South Wales, and the Dingo Farm at Chewton in Victoria.

The philosophy and attitude of such societies is admirable and their aims can be achieved if they take a united and scientifically correct approach. Otherwise the problems of traditional zoos become exacerbated and new problems can (and unfortunately do) emerge. Although there is growing evidence that distinct populations of dingoes exist in climatically distinct regions of Australia, the assumption that there are as many varieties of dingoes as there are regions is not only unproven but indicates the scientific naivety of those who assume this. Dingo characteristics should not have to conform (and be selectively bred) to such views — they are biased and scientifically unsound. It is important that people understand that natural selection has imposed a standard on dingoes, worldwide, over the past 6,000 years or so. The best available scientific sampling techniques must be used to ascertain the dingo's general and specific characteristics; these techniques draw on samples collected over most of the dingo's huge geographic range in Australia and Asia.

Other, more serious problems, are imminent. In mid 1993, through the efforts of the National Dingo Association and various preservation societies, the dingo was recognised by the Australian National Kennel Council as an official dog breed and adopted as Australia's national breed. Now that may open Pandora's box. Unless the registration of pure dingoes is done absolutely correctly, this landmark decision will in fact thwart the aims of these well-meaning societies and speed up the extinction of pure dingoes. How come? The following discussion first suggests the three crucial steps that, I believe, must be taken, and then suggests some of the consequences of not taking them.

Determining and registering pure dingoes

First, registered dingo breeders must ensure that their stocks comprise only pure dingoes. At present, this can only be done with skull measurements (see Appendix E) and backed up with scientifically confirmed coat colours and breeding patterns. Remember that even if external body characteristics seem to indicate a pure dingo, they are unreliable, and even the most experienced dingo experts can make mistakes by relying on them. For example, one famous 'dingo' named 'Brogo' turned out to be a hybrid after postmortem skull analysis, yet his offspring were featured as pure dingoes in several publications and films.

It may take 10 years or more before dingoes caught in the wild die a natural death so that their skulls can be examined, and proven dingo lines become registered. All offspring that turn out to be hybrids should then be discarded from the pure breeding line. In future, dingoes may be assessed for purity by DNA fingerprinting techniques (explained in Chapter 3) or by skull measurements from x-rays of live dingoes; but such techniques are presently not available.

The second important step is for the Australian National Kennel Council and its affiliations to register only pure dingoes — that is, animals whose parents are both from a pure dingo breeding line confirmed by the skull measurements of the founding parents of that line. Other dingo characteristics such as eye shape, stance and gait are of secondary importance. During the initial years when stocks of proven pure dingoes are low, care must be taken to minimise inbreeding and other genetic problems. It should be a registration requirement that the purity of every third generation of each breeding line is checked against skull measurements.

Informing the public and industry

The third important step is that the Australian National Kennel Council, other relevant societies and registered breeders should make considerable effort to educate the general public about the plight of dingoes and the measures being taken to preserve them. This education should include the responsibilities and problems of keeping dingoes in captivity and the reasons why most people would not qualify to keep a dingo. In addition, close collaboration with the RSPCA and other dog 'shelters' should be established to ensure that suspected dingoes and hybrids are not 'recycled' to members of the public.

More important, perhaps, is that considerable effort must also be made to win over members and supporters of the pastoral industry, particularly those in the sheep industry. Their beliefs are just as strong as those of the dingo lobby but their political pull is stronger, and rightly so, given the past and present contribution of pastoralism to the economy of Australia. Their concern is expressed, for instance, by South Australia's present government policy on the management of dingo populations. This states that the Animal and Plant Control Commission is

> ... gravely concerned by the vulnerability of the sheep industry to any relaxation of the effort in maintaining the sheep zone free of dingoes and actively supports the resolution of the Standing Committee on Agriculture (1975) that 'State and Territory authorities concerned should take action to ensure that dingoes or their offspring, pure bred or otherwise, are not kept in other than specially authorised Zoological Gardens or Circuses ... the Commission considers that the general possession, attempted domestication, or commercial exploitation of dingoes is unacceptable because such practices are likely to lead to the reintroduction of dingoes into sheep areas from which they have been removed, thus jeopardising the sheep industry in Australia.
>
> Similarly, the Commission also believes that the domestication or commercial exploitation by, for example, show breed societies, will undermine efforts to maintain the dingo essentially as a native wildlife species. Animal breeders are not able to breed for characteristics which maintain animals as a species in the wild — indeed the very nature of domestication means that wild characteristics will be actively selected against. Hence, domestication or commercialisation cannot be seen as an acceptable alternative to maintaining the species in the wild.

Thus a dilemma is apparent and dingoes will remain caught between the devil and the deep blue sea unless objective discussion ensues between participating groups and responsible governments.

THE PROBLEM WITH HYBRIDS

What is the attraction of keeping dingoes? Some of the reasons include participating in the quest to preserve the species; to keep them as pets; for oneupmanship (bragging); as a humane concern that animals should not be unnecessarily killed; and for financial gain. Most are worthwhile reasons, so what can go wrong? Lots, especially where hybrids are concerned! Unless the purity of dingoes is determined from a consistent, scientific basis, unscrupulous breeders may promote and pass off hybrids to unsuspecting people merely for financial gain. Unfortunately, profit is all that some people are concerned about, and this practice goes on in America where there are no requirements for laboratory tests to identify wolf-dog and coyote-dog hybrids. Dogs that are claimed to be the almost extinct American Indian dog (probably a primitive dingo — see Chapter 1) are being sold to unsuspecting people who believe they are participating in the noble cause of rescuing Indian dogs from extinction. Unfortunately most of the animals being sold are almost certainly coyote-husky hybrids. The red wolf, too, is classified as an endangered species, partly because it has hybridised with domestic dogs. There may be as many as 300,000 to 1.5 million wolf hybrids kept as pets in America, so there can be little hope of red wolves surviving either in the wild or in captivity.

Apart from the contamination of the dingo's gene pool, other problems exist with dingo-dog hybrids. For example, they pose more of a threat to the pastoral industry than pure dingoes do. Because hybrids can breed twice each year, they are capable of killing twice as many calves as pure dingoes can (see Chapter 9, Case 1). In urban areas hybrids are probably more dangerous to humans than most pure domestic dog breeds, at least if wolf hybrids are anything to go by. There have been eight human fatalities caused by 'pet' wolf hybrids in the past few years, and the California City Zoo terminated several experiments with wolf hybrids because more than 95% were too dangerous to handle.

Registered dingo breeders should not, therefore, expect financial gain for preserving the genetic stocks of dingoes — only cost recovery. After all, we should never lose sight of the fact that the only reason why dingoes were registered in the first place was to help preserve them so that if the bush ever becomes free of hybrids dingoes can be released there to determine their own destiny. There was never, and there never should be, any suggestion of (substantial) monetary gain for those engaged in this process.

Accordingly, the Australian National Kennel Council and associated bodies must provide the best option for governments to pass legislation that not only ensures the preservation of pure dingoes, but safeguards the credibility of preservation societies and combats the chicanery that unscrupulous 'dingo breeders' may employ. Some suggestions follow.

There should be only a few registered dingo breeders, perhaps only one from each state and territory so that designated procedures can be supervised. Registered dingoes should be ear-tattooed and other identifying characteristics described and photographed, so that when an animal dies it can be frozen in entirety until a government official or the Australian National Kennel Council confirms identification before removing the skull and measuring it. This should minimise the substitution of either dingo or hybrid skulls for the other, and curb the illegal tattooing of dingo-like hybrids.

Because dingoes do not make suitable pets or companion animals in most urban or rural situations, they should not be owned by members of the public except for those who have suitable facilities and are committed to preserving dingoes. In all such approved cases, the dingoes should be provided by registered breeders only, and all should be neutered once sperm or ova have been collected. This should minimise the risk that dingoes will become trendy pets for a time, then released to the streets or bush by owners who find them hard to handle or lose interest.

As already mentioned, dingo breeders cannot be expected to carry all the financial burden of preserving dingoes, but in the spirit of the movement nor should they expect more than cost recovery fees from the buyers of neutered dingoes. It seems reasonable that the pastoral industry and governments should give financial assistance to dingo breeders for maintaining dingo colonies and for sampling and preserving sperm and ova. There should also be licensing fees to pay for the enforcement of regulations by an appropriate authority such as the Australian National Kennel Council.

ARE ISLAND ZOOS THE GO?

In an ideal world people would own only neutered domestic dogs or dingoes; there would be adequate stocks of live dingoes and ova and sperm stored by dingo breeders, representing dingo genetic diversity from all major Australian (and Asian) habitats; the bush would be cleared of feral hybrids and feral dogs; stock would be totally protected from dingo predation; and the dream of releasing pure dingoes into the wild could then begin.

One day, perhaps, that dream may be realised. Meanwhile, the more pragmatic solution is to assume that most Australian habitats will never again be freely available to dingoes, no matter how successfully dingo breeders stockpile dingoes and their genes, and to focus on realistic options. For example, large offshore islands are perhaps the best hope of preserving dingoes in their natural habitat. There are many islands around the Australian coastline, and they represent many climates and habitats, except hot deserts. They are also big enough for dingoes to live and breed in partly or completely natural conditions.

Compared to most mainland dingo populations, the more isolated island populations probably contain higher proportions of pure dingoes. In any event, the nature and confines of an island would make it possible to minimise and eventually eliminate hybrids. For example, Fraser Island off the coast of Queensland has a large dingo population living in designated forest conservation areas. Whether or not these dingoes represent the purest strain of dingo in eastern Australia is irrelevant; the point is that the human residents recognise the tourist value of their dingoes and have taken steps to curb future contamination. For instance, a local law requires that all dogs owned by island residents must be neutered, and when the dogs die they cannot be replaced. Most of the residents are willing to comply not just because they value the tourist dollar but because they also take pride in preserving dingoes. This is vitally important, as before long Fraser Island will probably be one of the few locations where the Australian public and other tourists can see dingoes in a natural environment.

For this reason, another Fraser Island law that bans the hand-feeding of dingoes must be strictly adhered to; otherwise many dingoes will not only become dependent on food handouts but also accustomed *to the close presence of humans,* which increases the risk of dingoes annoying or biting humans. Dingoes are likely to do this to defend their food handouts or to protect breeding mates from apparent competitors (dingo and human), or as a displacement activity resulting from an altercation with other (more dominant) dingo group members. Nobody would ever wish to see a repeat of the tragic Chamberlain incident at Uluru (Ayers Rock) several years ago; yet, to my mind, that situation was set up by the practice of feeding dingoes so that some became used to approaching humans.

The purity of the Fraser Island dingo population must be assessed because many domestic dogs have lived there for many years, first along with sand miners and now with other residents. Such an assessment could easily be made now and monitored in the future by collecting and measuring the skulls of dingoes that die naturally or are culled by rangers.

Steps taken on Groote Eylandt, off the Northern Territory's coast, represent another preservation strategy in another habitat type. Although domestic dogs have been kept there by resident miners and local Aborigines for many years, previous surveys in the Top End and recent surveys on Groote Eylandt suggest that its dingo populations contain a very high proportion of pure animals. This situation is highly likely to continue and even improve because the BHP mining company is eliminating all feral animals, including hybrid dingoes, in accordance with its Native Fauna Conservation Program. This commitment, if supported by Aboriginal and non-Aboriginal residents, is likely to be one of the most significant steps in the preservation of dingoes in their natural environment, and may well turn out to be a role model.

Many other islands, such as Bathurst and Melville Islands, have dingo populations that make minimal contact with domestic dogs and these islands offer opportunities similar to those on Groote Eylandt. Yet other islands that have either no dingoes or hybrid populations that should be exterminated are suitable sites for releasing breeders' stock, preferably stock derived from dingo populations that will soon disappear from the mainland and, indeed, the world.

So what does the future hold for the dingo? In its travels throughout the world the dingo has faced many battles for survival against man and nature, from fullscale eradication campaigns and enormous fences to unjustified victimisation and subversive genetic manipulations. Although dingoes have won most of the battles, the cruel irony is that they are steadily losing the war, thanks to their evolutionary progeny, domestic dogs. In the end, their chances of continued survival in the wild will rest solely on the efforts of an informed public to stop contact between dingoes and domestic dogs, and to take pride in dingoes as native species whether they be Thai or Australian.

APPENDIX A

GLOSSARY OF ZOOLOGICAL AND BOTANICAL SPECIES

Mammals

African Hunting Dog	*Lycaon pictus*
Agile Wallaby	*Macropus agilis*
Antilopine Wallaro	*Macropus antilopinus*
Baboon	*Papio hamadryas*
Bilby (Greater)	*Macrotis lagotis*
Bilby (Lesser)	*Macrotis leucura*
Black Rat	*Rattus rattus*
Black-footed Tree-rat	*Mesembriomys gouldii*
Black-footed Rock Wallaby	*Petrogale lateralis*
Broad-toothed Rat	*Mastacomys fuscus*
Brown Antechinus	*Antechinus stuartii*
Brown Bear	*Ursus arctos*
Brown (Spinifex) Hopping Mouse	*Notomys alexis*
Brown Hyena	*Hyaena brunnea*
Brush-tailed Phascogale	*Phascogale tapoatafa*
Burrowing Bettong	*Bettongia lesueur*
Bush Rat	*Rattus fuscipes*
Cattle	*Bos taurus*
Central Hare-wallaby	*Lagorchestes asomatus*
Cheetah	*Acinonyx jubatus*
Coatis	*Nasua narica*
Common Brushtail Possum	*Trichosurus vulpecula*
Common Dunnart	*Sminthopsis murina*
Common Planigale	*Planigale maculata*
Common Ringtail Possum	*Pseudocheirus peregrinus*
Common Wombat	*Vombatus ursinus*
Coyote	*Canis latrans*
Crescent Nailtail Wallaby	*Onychogalea lunata*
Desert Bandicoot	*Perameles eremiana*
Desert Rat-kangaroo	*Caloprymnus campestris*
Dhole	*Cuon alpinus*
Dingo	*Canis lupus dingo* (*C. familiaris dingo*)
Domestic Dog	*Canis lupus familiaris* (*C. familiaris familiaris*)
Donkey	*Equus asinus*
Dusky Antechinus	*Antechinus swainsonii*
Dusky Rat	*Rattus colletti*
Dwarf Buffalo	*Anoas depressicornis*
Dwarf Mongoose	*Helogale parvula*
Eastern Barred Bandicoot	*Perameles gunnii*
Eastern Grey Kangaroo	*Macropus giganteus*
Echidna (Short-beaked)	*Tachyglossus aculeatus*
Euro (Common Wallaroo)	*Macropus robustus*
Fawn-footed Melomys	*Melomys cervinipes*
Fat-tailed Dunnart	*Sminthopsis crassicaudata*
Feral Cat	*Feliscatus (domesticus)*
Feral Cattle	*Bos taurus/Bos indicus*
Feral Horse	*Equus caballus*
Feral Pig	*Sus scrofa*
Feral Swamp (Water) Buffalo	*Bubalus bubalis*
Forrest's Short-tailed Mouse	*Leggadina forresti*
Golden Bandicoot	*Isoodon auratus*
Golden Jackal	*Canis aureus*
Greater Glider	*Petauroides volans*
Grizzly Bear	*Ursus horribilis arctos*
Horse	*Equus caballus*
House Mouse	*Mus domesticus (musculus)*
Hyena (Spotted)	*Crocuta crocuta*
Indian Wolf	*Canis lupus pallipes*
Koala	*Phascolarctos cinereus*
Lion	*Panthera leo*
Long-haired Rat	*Rattus villosissimus*
Long-nosed Bandicoot	*Perameles nasuta*
Long-nosed Potoroo	*Potorous tridactylus*
Mountain Brushtail Possum	*Trichosurus caninus*
Mountain Pygmy Possum	*Burramys parvus*
Northern Brown Bandicoot	*Isoodon macrourus*
Northern Brushtail Possum	*Trichosurus arnhemensis*
Northern Nail-tailed Wallaby	*Onychogalea unguifera*
Northern Quoll	*Dasyurus hallucatus*
Parma Wallaby	*Macropus parma*
Pig-footed Bandicoot	*Chaeropus ecaudatus*
Platypus	*Ornithorhynchus anatinus*
Polar Bear	*Ursus maritimus*
Puma	*Puma concolor*
Rabbit (European)	*Oryctolagus cuniculus*
Red Fox	*Vulpes vulpes*
Red Kangaroo	*Macropus rufus*
Red Wolf	*Canis lupus niger*
Red-necked Pademelon	*Thylogale thetis*
Red-necked Wallaby	*Macropus rufogriseus*
Rock Ringtail Possum	*Pseudocheirus dahli*
Rufous Hare-wallaby	*Lagorchestes hirsutus*
Sandy Inland Mouse	*Pseudomys hermannsbergensis*
Sheep	*Ovis aries*
Southern Brown Bandicoot	*Isoodon obesulus*
Spectacled Hare-wallaby	*Lagorchestes conspicillatus*
Spotted Hyena	*Crocuta crocuta*
Squirrel Glider	*Petaurus norfolcensis*
Sugar Glider	*Petaurus breviceps*

Swamp Rat	*Rattus lutreolus*
Swamp Wallaby	*Wallabia bicolor*
Tasmanian Devil	*Sarcophilus harrisii*
Thylacine	*Thylacinus cynocephalus*
Tiger	*Panthera tigris*
Water-rat	*Hydromys chrysogaster*
Wolf	*Canis lupus* spp.
Wurl-wurl (Kultarr)	*Antechinomys laniger*
Yellow-bellied Glider	*Petaurus australis*

Birds

Australian Bustard	*Ardeotis australis*
Australian Magpie	*Gymnorhina tibicen*
Australian Magpie Lark	*Grallina cyanoleuca*
Australian (Nankeen) Kestrel	*Falco cenchroides*
Barn Owl	*Tyto alba*
Barnacle Goose	*Branta leucopsis*
Black-faced Woodswallow	*Artamus cinereus*
Brown Songlark	*Cinclorhamphus cruralis*
Budgerigah	*Melopsittacus undulatus*
Bush Thicknee (Stone-curlew)	*Burhinus magnirostris (grallarius)*
Chestnut-rumped Thornbill	*Acanthiza uropygialis*
Cockatiel (Quarrion)	*Leptolophus (Nymphicus) hollandicus*
Cormorant	*Phalacrocorax* spp.
Crested Pigeon	*Geophaps (Ocyphaps) lophotes*
Crested Tern	*Sterna bergii*
Crow/Raven	*Corvus* sp.
Diamond Dove	*Geopelia cuneata*
Domestic Fowl	*Gallus gallus*
Emu	*Dromaius novaeholandiae*
Eurasian Coot	*Fulica atra*
Fairy Prion	*Pachyptila turtur*
Falcon	*Falco* sp.
Fulmar	*Fulmarus glacialoides*
Galah	*Cacatua roseicapilla*
King Parrot	*Alisterus scapularis*
Kookaburra	*Dacelo* sp.
Little Button-quail	*Turnix velox*
Little Corella	*Cacatua sanguinea (pastinator)*
Little Crow	*Corvus bennetti*
Little Penguin	*Eudyptula minor*
Magpie Goose	*Anseranas semipalmata*
Muttonbird (Shearwater)	*Puffinus (Puffinis)* sp.
Northern (White-cheeked) Rosella	*Platycercus venustus (eximius)*
Pied Butcherbird	*Cracticus nigrogularis*
Port Lincoln Ringneck	*Barnardius zonarius (barnardi)*
Rainbow Lorrikeet	*Trichoglossus haematodus*
Rajah (Burdekin) Shelduck	*Tadorna radjah*
Red-winged Parrot	*Aprosmictus erythropterus*
Richard's Pipit	*Anthus novaeseelandiae*
Spinifex (Plumed) Pigeon	*Petrophassa (Geophaps) plumifera*
Splendid Fairywren	*Malurus splendens*
Spotted Harrier	*Circus assimilis*
Spotted Nightjar	*Caprimulgus guttatus (argus)*
Superb Bluewren (Fairywren)	*Malurus cyaneus*
Superb Lyrebird	*Menura novaehollandiae*
Swamp Harrier	*Circus approximans*
Swan (Black)	*Cygnus atratus*
Wedge-tailed Eagle	*Aquila audax*
Welcome Swallow	*Hirundo neoxena*
Whistling Kite	*Haliastur sphenurus*
White-bellied Sea-eagle	*Haliaeetus leucogaster*
Whiteface	*Aphelocephala* sp.
White-winged Fairywren	*Malurus leucopterus*
Zebra Finch	*Taeniopygia (Peophila) guttata*

Reptiles

Bearded Dragon	*Pogona vitticeps (minor)*
Bobtail (Shingleback) Skink	*Tiliqua (Trachydosaurus) rugosa*
Central Netted (Ground) Dragon	*Ctenophorus nuchalis*
Earless (Lined) Dragon	*Tympanocryptis lineata*
Frilled Lizard	*Chlamydosaurus kingii*
Gilbert's Dragon (Lashtail)	*Gemmatophora (Amphibolurus) gilberti*
King Brown (Mulga) Snake	*Pseudechis australis*
Lace Monitor	*Varanus varius*
Leonhardi's (Ctenotus) Skink	*Ctenotus leonhardii*
Little Spotted Snake	*Rhinoplocephalus punctatus*
Long-nosed Water Dragon (Lashtail)	*Gemmatophora (Amphibolurus) longirostris*
Military (Sand) Dragon	*Ctenophorus isolepis*
Painted (Ground) Dragon	*Ctenophorus pictus*
Panoptes (Yellow-spotted) Monitor	*Varanus panoptes*

Perentie	*Varanus giganteus*
Python	Boidae sp.
Ridge-tailed Monitor (Ocellate)	*Varanus acanthurus*
Sand Monitor	*Varanus gouldii*
Short-tailed (Pygmy) Monitor	*Varanus brevicauda*
Spencer's Monitor	*Varanus spenceri*
Spiny-tailed (Northern) Gecko	*Diplodactylus ciliarus*
Whip Snake	*Demansia* sp.

Insects

Beetle	*Coleoptera*
Caterpillar	*Lepidoptera*
Cricket	*Grylloidea*
Emperor Gum Moth	*Antheraea eucalyptii*
Grasshopper	*Orthoptera*
Praying Mantis	*Mantodea*

Botanical species

Chaenopods (saltbush)	*Atriplex* spp.
Flinders Grass	*Iseilema membranacea*
Mitchell Grass	*Astrebla pectinata*
Mulga	*Acacia aneura*
Succulent Berry	*Diospyros calycantha*

APPENDIX B

ESTIMATING THE AGE OF DINGOES

This method estimates the age of dingoes (in days) by head-length (mm) and eye-lens weight (mg). Data derive from studies by Catling et al. (1991) of known-age captive dingoes. Head-length is measured from the back of the sagittal crest to the tip of the nose of live dingoes. The eye-lenses of dead dingoes were first fixed in 10% formalin, then dried at 82°C for 96 hours before weighing.

Age of Females (d)	Age of Males (d)	Character
		Head-length (mm)
4	4	65
8	7	70
11	10	75
15	14	80
19	17	85
23	21	90
27	25	95
31	29	100
36	33	105
41	37	110
46	42	115
52	47	120
57	52	125
64	58	130
71	63	135
78	70	140
86	76	145
96	84	150
106	91	155
117	100	160
131	110	165
147	120	170
		Dried eye-lens weight (mg)
160	152	130
178	169	140
200	188	150
226	211	160
256	238	170
297	271	180
346	312	190
412	364	200
501	432	210

THE DIET OF DINGOES IN SIX MAJOR HABITATS OF AUSTRALIA

The diet of dingoes (% occurrence) in six major habitats of Australia as indicated from the remains of prey in 12,802 stomachs and faeces collected between 1966–86. Study site numbers refer to sites detailed in Figure 2.2 and Table 2.1. The scientific names of prey species are given in Appendix A. Prey size is defined in Box 7.1. Despite the immense range of potential prey species across Australia, only ten species formed almost 80% of the diet; thus in terms of dietary intake, dingoes are specialists rather than generalists. However, dingoes use a large range of hunting tactics to catch prey and are generalist predators in this respect.

	Wet–Dry Tropics North Australia		Arid & Semi-arid Central Australia		Arid South-West Australia
No. Samples	6722 faeces		1480 stomachs		131 faeces/stomachs
Study Sites	Site 1		Sites 2,4,5,6		Site 7
Years of Study	1980–86		1966–75		1982–86
All Mammals	73.4		106.1		109.2
Large Mammals	12.5		36.4		39.7
Medium Mammals	26.6		41.7		69.5
Small Mammals	34.3		28.0		0
Reptiles	0.1		14.1		1.5
Birds	33.8		11.9		2.3
Insects	1.3		4.1		0.8
Vegetation	7.3		0.1		0
Others	0.4		0.5		0.8
Dusky Rat	33.9	Rabbit	37.9	Rabbit	63.4
Magpie Goose	32.5	Cattle	23.3	Red Kangaroo	32.1
Agile Wallaby	15.1	Long-haired Rat	17.6	Cattle	7.6
Northern Brushtail Possum	9.7	Red Kangaroo	10.2	Red Fox	3.8
Grass (undetermined spp.)	7.1	Central Netted Dragon	7.8	Little Crow	2.3
Feral Water Buffalo	5.8	Undetermined Small Mammal	3.8	Bobtail Skink	1.5
Feral Pig	3.5	House Mouse	3.6	Feral Cat	1.5
Unidentified Matter	2.6	Grasshopper	2.7	Centipede/Millipede	0.8
Antilopine Wallaroo	1.8	Bearded Dragon	2.2	Dingo	0.8
Northern Brown Bandicoot	1.4	Zebra Finch	2.1	Grasshopper	0.8
Feral Cattle	1.3	Undetermined Bird	2.0		
Bird (undetermined spp.)	0.7	Feral Cat	1.8		
Insect (undetermined spp.)	0.7	Galah	1.8		
Beetle	0.4	Budgerigah	1.5		
Undetermined Small Mammal	0.3	Brown Hopping Mouse	1.4		
Egg (mostly Magpie Goose)	0.2	Crow/Raven	1.1		
Feral Horse	0.2	Euro	1.1		
Grasshopper	0.2	Sand Monitor	1.1		
Northern Quoll	0.2	Sandy Inland Mouse	1.1		
Succulent Berry	0.2	Undetermined Reptile	0.9		
Bush Thicknee	0.1	Beetle	0.8		
Dove (undetermined spp.)	0.1	Horse	0.7		
Feral Cat	0.1	Northern Nail-tail Wallaby	0.7		
Northern Rosella	0.1	Spectacled Hare-wallaby	0.7		
Rajah Shelduck	0.1	Undetermined Macropod	0.7		

North Australia continued

Undetermined Snake	0.1
Australian Kestrel	<0.1
Australian Magpie Lark	<0.1
Black-footed Tree-rat	<0.1
Cockatiel	<0.1
Common Planigale	<0.1
Crab (undetermined sp.)	<0.1
Galah	<0.1
Kingfisher (undetermined sp.)	<0.1
Little Corella	<0.1
Owl (undetermined sp.)	<0.1
Pigeon (undetermined sp.)	<0.1
Rainbow Lorikeet	<0.1
Red-winged Parrot	<0.1
Sand Monitor/Panoptes Monitor	<0.1
Undetermined Lizard	<0.1

Central Australia continued

Australian Bustard	0.5
Fat/Meat/Bone (undet. spp.)	0.5
Spencer's Monitor	0.5
Crested Pigeon	0.4
Ridge-tailed Monitor	0.4
Donkey	0.3
Echidna	0.3
Emu	0.3
Forrest's Mouse	0.3
Leonhard's Skink	0.3
Undetermined Insect	0.3
Whistling Kite	0.3
Diamond Dove	0.2
Spotted Nightjar	0.2
Whip Snake (undet. sp.)	0.2
Whiteface	0.2
Wurl-wurl	0.2
Australian Magpie	0.1
Barn Owl	0.1
Berries (undetermined sp.)	0.1
Bilby	0.1
Black-faced Woodswallow	0.1
Brown Songlark	0.1
Caterpillar	0.1
Chestnut-rumped Thornbill	0.1
Cricket	0.1
Domestic Chicken	0.1
Earless Dragon	0.1
Falcon (undetermined sp.)	0.1
Fat-tailed Dunnart	0.1
Gilbert's Dragon	0.1
King Brown Snake	0.1
Little Spotted Snake	0.1
Little Button-Quail	0.1
Long-nosed Water Dragon	0.1
Military Dragon	0.1
Painted Dragon	0.1
Perentie	0.1
Pied Butcherbird	0.1
Port Lincoln Ringneck	0.1
Praying Mantis	0.1
Python (undetermined sp.)	0.1
Red Fox	0.1
Richard's Pipit	0.1
Sheep	0.1
Short-tailed Monitor	0.1
Spinifex Pigeon	0.1
Spiny-tailed Gecko	0.1
Splendid Fairy Wren	0.1
Welcome Swallow	0.1
White-winged Fairy Wren	0.1

	Semi-Arid North-West Australia	**Cool Coastal Mountains Southeast Australia**	**Humid Coastal Mountains East Australia**
	413 faeces/stomachs Site 3 1977-84	2063 faeces/stomachs Sites 9,10,11 1966-80	1993 faeces Site 8 1969-74
All Mammals	105.1	108.7	106.8
Large Mammals	100.0	22.9	0.4
Medium Mammals	4.8	72.6	85.8
Small Mammals	0.2	13.2	20.6
Reptiles	3.4	1.0	1.3
Birds	5.6	21.7	2.7
Insects	2.9	2.2	<0.1
Vegetation	0	1.2	0.2
Others	0.2	16.3	1.4

Semi-Arid North-West Australia		Cool Coastal Mountains Southeast Australia		Humid Coastal Mountains East Australia	
Red Kangaroo/Euro	80.6	Swamp Wallaby	17.9	Swamp Wallaby	30.5
Cattle	11.4	Wallaby (undetermined spp.)	15.8	Bush Rat	12.2
Sheep	8.0	Wombat	15.0	Red-necked Wallaby	11.1
Bird (undetermined spp.)	5.6	Bone/Meat/Fat/Hide (undetermined spp.)	11.5	Brushtail Possums (Common, Mountain)	6.9
Reptile (undetermined spp.)	3.4	Rabbit	10.5	Bandicoots (Long-nosed, Southern Brown)	6.8
Insects (undetermined spp.)	2.9	Common Ringtail Possum	8.0	Rabbit	6.4
Echidna	2.2	Waterbird (undetermined spp.)	7.7	Antechinuses (Brown, Dusky)	5.8
Dingo	1.7	Red-necked Wallaby	5.3	Parma Wallaby	4.6
Feral Cat	0.5	Possum (undetermined spp.)	5.1	Common Ringtail Possum	4.4
Bat (undetermined sp.)	0.2	Undetermined Rattus spp.	5.0	Red-necked Pademelon	3.8
Fish (undetermined sp.)	0.2	Little Penguin	4.4	Echidna	3.5
Red Fox	0.2	Fish (undetermined spp.)	3.7	Long-nosed Potoroo	1.7
Rothschild's Rock-wallaby	0.2	Mutton Bird	3.6	Greater Glider	1.5
		Echidna	3.3	Eastern Grey Kangaroo	0.3
		Antechinuses (Brown, Dusky)	3.0	Cattle	<0.1
		Eastern Grey Kangaroo	2.8	Other Small Mammals (Swamp Rat, Water Rat, Common Dunnart, Fawn-footed Melomys, Brush-tailed Phascogale, House Mouse)	2.5
		Bush Rat	2.4	Other Mammals (10 spp. including Euro, Southern Sheep, Platypus, Sugar Glider, Squirrel Glider, Yellow-bellied Glider)	4.6
		Common Brushtail Possum	2.3	Birds (including Kookaburra, Cockatoo and Parrots)	2.7
		Eurasian Coot	2.3	Reptiles (undetermined spp.)	1.3
		Bandicoots (Long-nosed, Brown)	2.2	Others (5 Classes, mainly Insects & Crustacea)	1.4
		Large Mammal (undetermined spp.)	2.0		
		Bird (undetermined spp.)	1.6		
		Swan	1.6		
		Insect (undetermined spp.)	1.5		
		Sheep	1.4		
		House Mouse	1.2		
		Solanum Seeds	1.2		
		Crab (undetermined spp.)	1.1		
		Swamp Rat	0.9		
		Feral Pig	0.8		
		Lizard (undetermined spp.)	0.8		

Southeast Australia continued

Cattle	0.7
Long-nosed Potoroo	0.6
Beetles	0.5
Red Fox	0.4
Broad-toothed Rat	0.3
Feral Cat	0.3
Koala	0.3
Medium-sized Mammal (undetermined sp.)	0.3
Horse	0.2
Black Rat	0.1
Crested Tern	0.1
Domestic Chicken	0.1
Grasshoppers	0.1
Greater Glider	0.1
Lace Monitor	0.1
Mountain Brushtail Possum	0.1
Small Mammal (undetermined spp.)	0.1
Snake (undetermined spp.)	0.1
Yellow-bellied Glider	0.1
Cormorant	<0.1
Emperor Gum Moth Pupae	<0.1
Emu	<0.1
Fairy Prion	<0.1
Frog (undetermined spp.)	<0.1
King Parrot	<0.1
Superb Lyrebird	<0.1
Mountain Pygmy Possum	<0.1
Sugar Glider	<0.1
Superb Blue Wren	<0.1
Water Rat	<0.1

APPENDIX D

PARASITES AND PATHOGENS RECORDED FROM DINGOES IN AUSTRALIA

Organism	Locality recorded	Incidence (% dingoes sampled)	Highest burden (n)	Site of infection	Likely effect on dingo
Cestodes					
Tapeworms					
Echinococcus granulosus	Queensland	90	78940	Small intestine	Non-pathogenic
	SE Highlands	62	>80000		
Taenia pisiformis	Central Australia	97	>20	Small intestine	Loss of condition if heavy burden
	SE Highlands	46	39		
Taenia hydatigena	SE Highlands	23	8	Small intestine	Loss of condition if heavy burden
Taenia serialis	SE Highlands	23	280	Small intestine	Loss of condition if heavy burden
Dipylidium caninum	SE Highlands	2	277	Small intestine	Unknown
	Queensland	Present	?		
Spirometra erinacei	SE Highlands	38	23	Small intestine	Unknown
	Queensland	Present	?		
Undetermined species	Central Australia	Low	?	Small intestine	Unknown
	North Australia	Low	?		
	Barkly Tableland	28	?		
Nematodes					
Hookworms					
Uncinaria stenocephala	SE Highlands	50	>200	Small intestine	Severe anaemia, often fatal
Ancylostoma caninum	SE Highlands	Low	?	Small intestine	Severe anaemia, often fatal
	Queensland	Present	?		
Undetermined species	North Australia	Medium	?	Stomach/intestine	Unknown
Roundworms					
Toxacara canis	SE Highlands	14	11	Small intestine	Anaemia, diarrhoea, loss of condition, larval stage in lungs may cause pneumonia
Undetermined species	Central Australia	2	?	Small intestine	Unknown
	North Australia	Low	?		
Heartworm					
Dirofilaria immitis	North Australia	>90 in adults	>200	Right ventricle & pulmonary artery	Circulatory difficulty, fatal with heavy burden
	Barkly Tableland	5	3		
	SE Highlands	<1	?		
Lungworm					
Oslerus osleri	SE Highlands	83	63	Trachea, most at bifurcation of bronchi	Coughing, difficulty in breathing, possibly fatal in pups
	Central Australia	6	?		
Whipworm					
Trichurus vulpis	SE Highlands	1	16	Caecum & large intestine	Diarrhoea, severe anaemia, probably fatal in pups
Spiruroid					
Cyathospirura dasyuridis	SE Highlands	2	>200	Alimentary tract	Unknown
Thorn-headed worm					
Acanthacephala sp.	Barkly Tableland	Low	1	Stomach	Unknown
Viruses					
Canine distemper					
Paramyxovirus	Central Australia	Large epizootics occur about 1 in 10 years		Respiratory tract	Usually fatal
	Barkly Tableland				
Canine Hepatitis					
Adenovirus	Central Australia	Recorded in captive colony		Liver	Suppressed breeding in females
	Barkly Tableland	Present			Unknown, probably only fatal in pups

Organism	Locality recorded	Incidence (% dingoes sampled)	Highest burden	Site of infection (n)	Likely effect on dingo
Protozoa					
Coccidiosis					
Isospora rivolta	SE highlands	Present		Alimentary tract	Diarrhoea, probably fatal in young pups
Eimeria canis	SE highlands	Present			
Sarcosporidiosis					
Sarcocystis sp.	Queensland	Common		Striated muscle & heart muscle	Non-pathogenic
Insects					
Biting lice					
Trichodectes canis	South-east Australia	Present		Skin	Severe irritation, possibly anaemia & death in pups
Undetermined species	Barkly Tableland	7	?	Skin	Unknown
	Central Australia	2	?		
Fleas					
Ctenocephalides canis	Probably widespread	Present	?	Skin	Dermatitis, loss of condition, host for heartworm and tapeworm (*Dipylidium caninum*)
Echidnophaga myrmecobii	Central Australia	21	?	Skin	Unknown
	Barkly Tableland	5	?		
Undetermined species	North Australia	Low	?	Skin	Unknown
Marchflies (Tabanidae)	Widespread			Skin	Possibly debilitating for pups & emaciated adults
Mosquitoes (Culicidae)	Widespread			Skin	Some species host heartworm
Blowflies (Calliphoridae)	Widespread			Wounds	Possibly debilitating for injured animals
Ticks					
Ixodes holocyclus	SE coastal highlands	24	?	Skin	Irritation, oedema, paralysis, probably fatal in pups
Rhipicephalus sanguineus	Central Australia	2	?	Skin	Irritation, loss of condition, anaemia, possible vector for canine babesiosis
	North Australia	Low	?		
Amblyomma triguttatum	Queensland	Present	?	Skin	Unknown
Mites					
Sarcoptic mange					
Sarcoptes scabiei	Western Australia	20		In skin	Progressive loss of hair & emaciation if mange is extensive
	North Australia	5			
	SE highlands	2			
	Central Australia	1			
	Queensland	Present			
Demodectic mange					
Demodex folliculorum	Central Australia	Recorded in captive colony		Sebaceous glands & skin follicles around eyes & nose	Alopecia & emaciation
Pentastome Arthropod					
Tongueworm					
Linguatula serrata	SE Australia	<1	?	Nasal cavities	Unknown, probably non-pathogenic
Fungus					
Ringworm					
Microsporon canis	Central Australia	Recorded in captive colony		Skin	Non-pathogenic
Leech					
Hirudo sp.	SE highlands	<1	?	Skin	Non-pathogenic

Identifying dingo, dog and hybrid skulls

To classify an 'unknown' canid skull as a pure dingo, domestic dog or hybrid with a 95% level of probability, the following steps are taken:

Step 1 Referring to Figure on page 190 for an explanation and location of anatomical terms, take eight straight-line measurements (x_1 to x_8), preferably to two decimal places, using vernier calipers.

Step 2 Substitute the values (x_1 to x_8) in the following equation to determine a composite skull value (Y).

$$Y = 0.249x_1 - 0.261x_2 + 1.999x_3 - 1.137x_4 + 0.318x_5 + 0.475x_6 - 0.205x_7 + 0.136x_8 - 3.717$$

Step 3

If the calculated Y value is:	the 'unknown' skull is:
$\leq$ -1.394	Domestic dog
-1.393 to 1.270	Hybrid
$\geq$ 1.271	Dingo

Note: Sometimes the composite score in Step 3 will suggest that the skull is dingo even though one or more of the individual measurements (Step 1) are outside the dingo range (given in the table below (CL = confidence limit, se = standard error of the mean).

	Dingo		Hybrid	Dog
	Mean ± se	95% CL	Mean ± se	Mean ± se
x_1	25.1 ± 0.0	22.8–27.4	22.1 ± 0.3	20.8 ± 0.3
x_2	60.3 ± 0.4	56.8–63.8	60.1 ± 0.5	62.8 ± 0.7
x_3	7.5 ± 0.1	6.9– 8.2	6.8 ± 0.7	6.8 ± 0.1
x_4	9.5 ± 0.1	8.6–10.5	9.4 ± 0.1	9.8 ± 0.1
x_5	33.5 ± 0.3	30.0–37.0	30.3 ± 0.4	28.4 ± 0.4
x_6	11.6 ± 0.1	10.2–13.0	10.7 ± 0.2	10.2 ± 0.2
x_7	55.9 ± 0.3	52.4–59.4	55.2 ± 0.5	58.2 ± 0.6
x_8	54.6 ± 0.3	50.8–58.4	49.8 ± 0.7	50.5 ± 0.6

Step 4 Thus the most confident assessment that an 'unknown' animal is pure dingo will include positive results to the following criteria:

1 The composite measurement Y (Step 2) occurs in the dingo range (Step 3).

2 All eight individual skull measurements x_1 to x_8 (Step 1) occur in the dingo range (table above).

3 The coat colour is ginger, black-and-tan, black, or white, with no specking in white markings.

4 Females exhibit a seasonal breeding pattern.

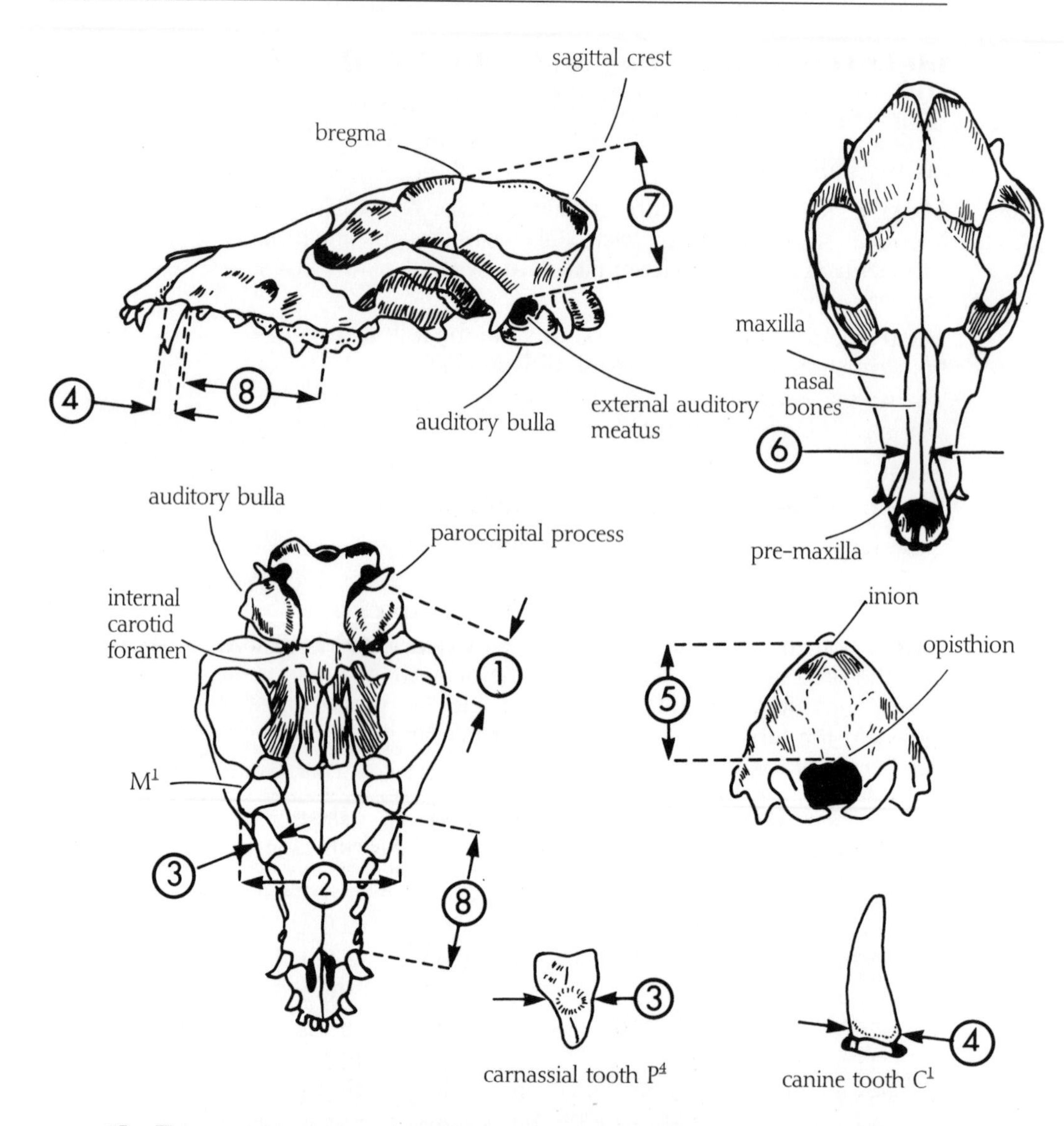

Skull measurements. Numbers in circles refer to x_1 to x_8 below.

x_1 = length of auditory bulla (measured from where it abuts the paroccipital process to the internal carotid foramen, excluding any projection on the foramen).
x_2 = maximum maxillary width (measured at about the junction of the P^4 and M^1 teeth).
x_3 = mid-crown width of the P^4 tooth (measured through the highest cusp in a lateral direction)
x_4 = basal crown length of C^1 (measured along the tooth row).
x_5 = opisthion to inion (measured from a central inion point and not including the notch in the opisthion, if present).
x_6 = width of both nasal bones (measured at premaxilla-maxilla suture).
x_7 = cranial height (measured from the upper notch of the external auditory meatus to the bregma, including the sagittal crest).
x_8 = distance between the posterior alveolar rims of C^1 – P^4.

CHAPTER 1
Origin, ancestry and world distribution

Kodansha Encyclopedia of Japan (1983). Japanese dogs. (Kodansha Ltd: Tokyo.) pp. 124–6.

Clutton-Brock, J. (ed.) (1989). *The Walking Larder: Patterns of Domestication, Pastoralism and Predation.* (Unwin Hyman: London.)

Clutton-Brock, J. (1992). 'The process of domestication'. *Mammal Review* 22: 79–85.

Corbett, L.K. (1985). 'Morphological comparisons of Australian and Thai dingoes: a reappraisal of dingo status, distribution and ancestry'. *Proceedings of the Ecological Society of Australia* 13: 277–91.

Finlayson, H.H. (1935). *The Red Centre: Man and Beast in the Heart of Australia.* (Angus and Robertson: Sydney.)

Flamholtz, C.J. (1986). *A Celebration of Rare Breeds, Vol. I. The Canaan Dog; The Telomian Dog.* (OTR Publications: Alabama.)

Flamholtz, C.J. (1991). *A Celebration of Rare Breeds. Vol. II. The New Guinea Singing Dog* (OTR Publications: Alabama.)

Harney, W.E. (1951). 'The dingo'. *Walkabout*, 1 July, pp. 36–7.

Lumholtz, C. (1889). *Amongst Cannibals.* (J. Murray: London.)

Manwell, C. & Baker, C.M.A. (1984). 'Domestication of the dog: hunter, food, bed-warmer, or emotional object?' *Sonderdruck aus Zeitschrift fur Tierzuchtung und Zuchtungsbiologie* 101: 241–56.

Olsen, S.J. (1985). *Origins of the Domestic Dog: The Fossil Record.* (University of Arizona Press: Tucson.)

Olsen, S.J., & Olsen, J.W. (1977). 'The Chinese wolf: ancestor of New World dogs'. *Science* 197: 533–5.

Pickering, M. (1992). 'Garawa methods of game hunting, preparation and cooking'. *Records of the South Australian Museum* 26: 9–23.

Thomson, D.F. (1949). 'Arnhem Land: explorations among an unknown people. Part III. On foot across Arnhem Land'. *Geographical Journal* 114: 53–67.

Thorne, A. & Raymond, R. (1989). *Man on the Rim: The Peopling of the Pacific* (Angus & Robertson: Sydney.)

CHAPTER 2
Studying dingoes

Brunner, H. & Coman, B.J. (1974). *The Identification of Mammalian Hair.* (Inkata Press: Melbourne.)

Corbett, L.K. (1974). 'Contributions to the biology of dingoes (Carnivora: Canidae) in Victoria'. Unpub. Master of Science Thesis, Monash University.

Corbett, L.K. (1989). 'Assessing the diet of dingoes from feces: a comparison of 3 methods'. *Journal of Wildlife Management* 53: 343–6.

Green, B. (1976). 'The use of etorphine hydrochloride (M99) in the capture and immobilization of wild dingoes (*Canis familiaris dingo*)'. *Australian Wildlife Research* 3: 123–8.

Green, B. (1978). 'Estimation of food consumption in the dingo, *Canis familiaris dingo*, by means of ^{22}Na turnover'. *Ecology* 59: 207–10.

Larter, N.C., Arcese, P., Rajamahendran, R. & Gates, C.C. (1993). 'Measurement of immunoreactive progestins excreted in faeces as a potential indicator of pregnancy'. *Wildlife Research* 20: 739–43.

Putman, R.J. (1984). 'Facts from faeces'. *Mammal Review* 14: 79–97.

Thomson, P.C. (1992). 'Capture of dingoes with tranquilliser darts loaded with ketamine hydrochloride and xylazine hydrochloride'. *Wildlife Research* 19: 601–3.

Thomson, P.C. & Rose, K. (1992). 'Age determination of dingoes from characteristics of canine teeth'. *Wildlife Research* 19: 597–9.

CHAPTER 3
Characteristics and identity

Burns, M. & Fraser, M.N. (1966). *Genetics of the Dog.* (2nd edn) (Oliver & Boyd: Edinburgh.)

Catling, P.C. (1979). 'Seasonal variation in plasma testosterone and the testis in captive male dingoes, *Canis familiaris dingo*'. *Australian Journal of Zoology* 27: 939–44.

Catling, P.C., Corbett, L.K. & Newsome, A.E. (1992). 'Reproduction in captive and wild dingoes (*Canis familiaris dingo*) in temperate and arid environments of Australia'. *Wildlife Research* 19: 195–205.

Catling, P.C., Corbett, L.K. & Westcott, M. (1991). 'Age determination in the dingo and crossbreeds'. *Wildlife Research* 18: 75–83.

Cole, S.R., Baverstock, P.R. & Green, B. (1977). 'Lack of genetic differentiation between domestic dogs and dingoes at a further 16 loci'. *The Australian Journal of Experimental Biology and Medical Science* 55: 229–32.

Corbett, L.K. (1974). Contributions to the biology of dingoes (Carnivora: Canidae) in Victoria, op. cit., Ch. 2.

Corbett, L.K. (1985). 'Morphological comparisons of Australian and Thai dingoes: a reappraisal of dingo status, distribution and ancestry', op. cit., Ch. 1.

Corbett, L.K. (1988). 'The identity and biology of Thai dingoes'. Unpub. Final Report to the National Research Council of Thailand.

Frith, H.J. & Calaby, J.H. (eds) (1974). 'Fauna survey of the Port Essington District, Cobourg Peninsula, Northern Territory of Australia'. *CSIRO Division of Wildlife Research Technical Paper* No. 28, pp. 1–208.

Ginsberg, J.R. & Macdonald, D.W. (1990). *Foxes, Wolves, Jackals, and Dogs. An Action Plan for the Conservation of Canids. IUCN/SSC Canid and Wolf Specialist Groups.* (IUCN Publications Services Unit: Cambridge.)

Honacki, J.H., Kinman, K.E. & Koeppl, J.W. (eds) (1982). *Mammal Species of the World: A Taxonomic and Geographic Reference.* (Allen Press and Assn of Systematic Collections: Lawrence, Kansas.)

Iredale, T. (1947). 'The scientific name of the dingo'. *Proceedings of the Royal Society of New South Wales.* pp 35–6.

Jones, E. (1990). 'Physical characteristics and taxonomic status of wild canids, *Canis familiaris*, from the eastern highlands of Victoria'. *Australian Wildlife Research* 17: 69–81.

Jones, E. & Stevens, P.L. (1988). 'Reproduction in wild canids, *Canis familiaris*, from the eastern highlands of Victoria'. *Australian Wildlife Research* 15: 385–94.

Little, C.C. (1957). *The Inheritance of Coat Colour in Dogs.* (Cornell University Press: Ithaca New York.)

Lowrie, P. & Wells, S. (1991). 'Genetic fingerprinting'. *New Scientist* 52: 1–4.

Newsome, A.E., Corbett, L.K. & Carpenter, S.M. (1980). 'The identity of the dingo. I. Morphological discriminants of dingo and dog skulls'. *Australian Journal of Zoology* 28: 615–25.

Newsome, A.E. & Corbett, L.K. (1982). 'The identity of the dingo. II. Hybridisation with domestic dogs in captivity and in the wild'. *Australian Journal of Zoology* 30: 365–74.

Newsome, A.E. & Corbett, L.K. (1985). 'The identity of the dingo. III. The incidence of dingoes, dogs and hybrids and their coat colours in remote and settled regions of Australia'. *Australian Journal of Zoology* 33: 363–75.

Oppenheimer, E.C. & Oppenheimer, J.R. (1975). 'Certain behavioural features in the pariah dog (*Canis familiaris*) in West Bengal'. *Applied Annuals of Ethology* 2: 1–11.

Roy, M.S., Geffen, E., Smith, D., Ostrander, E. & Wayne, R.K. (1994). 'Patterns of differentiation and hybridisation in North American wolf-like canids revealed by analysis of microsatellite loci'. *Molecular Biology and Evolution* (in press).

Sheldon, J.W. (1992). *Wild Dogs: The Natural History of the Nondomestic Canidae.* (Academic Press, Inc: San Diego.)

Spencer, B. (ed.) (1896). *Report of the Work of the Horn Expedition to Central Australia. Part II. Zoology.* (Melville, Mullen and Slade: Melbourne.)

Thomson, P.C. (1992). 'The behavioural ecology of dingoes in north-western Australia. I. The Fortescue River study area and details of captured dingoes'. *Wildlife Research* 19: 509–18.

Thomson, P.C., Rose, K. & Kok, N.E. (1992). 'The behavioural ecology of dingoes in north-western Australia. V. Population dynamics and variation in the social system'. *Wildlife Research* 19: 565–84.

Wayne, R.K., Lehman, N., Allard, M.W. & Honeycutt, R.L. (1992). 'Mitochondrial DNA variability of the grey wolf: genetic consequences of population decline and habitat fragmentation'. *Conservation Biology* 6: 559–69.

CHAPTER 4
Living areas and movements

Best, L.W. (1978). 'Dingo movements in central Australia'. *Handbook of Working Papers. Australian Vertebrate Pest Control Conference, Canberra 1978,* pp. 38–40.

Best, L.W. (nd). 'Movements and home range of dingoes (*Canis familiaris dingo*) in central Australia'. Unpub. ms.

Catling, P.C. (1978). 'Dingo movements in south-eastern New South Wales'. *Handbook of Working Papers. Australian Vertebrate Pest Control Conference, Canberra 1978,* pp. 40–2.

Harden, R.H. (1985). 'The ecology of the dingo in north-eastern New South Wales. I. Movements and home range'. *Australian Wildlife Research* 12: 25–37.

McIlroy, J.C., Cooper, R.J., Gifford, E.J., Green, B.F. & Newgrain, K.W. (1986). 'The effect on wild dogs, *Canis f. familiaris*, of 1080-poisoning campaigns in Kosciusko National Park, NSW'. *Australian Wildlife Research* 13: 535–44.

Newsome, A.E., Catling, P.C. & Corbett, L.K. (1983). 'The feeding ecology of the dingo. II. Dietary and numerical relationships with fluctuating prey populations in south-eastern Australia'. *Australian Journal of Ecology* 8: 345–66.

Sheldon, J.W. (1992). *Wild Dogs: The Natural History of the Nondomestic Canidae,* op. cit., Ch. 3.

Thomson, P.C. (1992). 'The behavioural ecology of dingoes in north-western Australia. IV. Social and spatial organisation, and movements'. *Wildlife Research* 19: 543–63.

Whitehouse, S.J.O. (1977). 'Movements of dingoes in Western Australia'. *Journal of Wildlife Management* 41: 575–6.

CHAPTER 5
Behaviour and communication

Corbett, L. & Newsome, A. (1975). 'Dingo society and its maintenance: a preliminary analysis'. *The Wild Canids: their Systematics, Behavioural Ecology and Evolution.* (ed. Fox, M.W.), (Van Nostrand Reinhold Company: New York.) pp. 369–79.

Ortolani, A. (1990). 'Howling vocalisations of wild and domestic dogs: a comparative behavioural and anatomical study'. BADissertation, Hampshire College USA.

Schenkel, R. (1967). 'Submission: its features and function in the wolf and the dog'. *American Zoologist* 7: 319–29.
Schenkel, R. (1947). 'Expression studies of wolves'. *Behaviour* 1: 81–129.
Sheldon, J.W. (1992), op. cit., Ch. 3.

CHAPTER 6
Social dynamics

Corbett, L.K. (1988). 'Social dynamics of a captive dingo pack: population regulation by dominant female infanticide'. *Ethology* 78: 177–98.
Harden, R.H. (1985). 'The ecology of the dingo in north-eastern New South Wales. I. Movements and home range', op. cit., Ch. 4.
McIlroy, J.C. et al. (1986). 'The effect on wild dogs, *Canis f. familiaris*, of 1080-poisoning campaigns in Kosciusko National Park', NSW, op. cit., Ch. 4.
Newsome, A.E. et al. (1983). 'The feeding ecology of the dingo. II', op. cit., Ch. 4.
Sheldon, J.W. (1992), op. cit., Ch. 3.
Thomson, P.C. (1992). 'The behavioural ecology of dingoes in north-western Australia. II. Activity patterns, breeding season and pup rearing'. *Wildlife Research* 19: 519–30.
Thomson, P.C. (1992). 'The behavioural ecology of dingoes in north-western Australia. IV', op. cit., Ch. 4.

CHAPTER 7
Feeding ecology

Begon, M. & Mortimer, M. (1986). *Population Ecology. A Unified Study of Animals and Plants.* (2nd edn) (Blackwell Scientific Publications: Oxford.)
Castairs, J.L. (1974). 'The distribution of *Rattus villosissimus* (Waite) during plague and non-plague years'. *Australian Wildlife Research* 1: 95–106.
Coman, B.J. (1972). 'Helminth parasites of the dingo and feral dog in Victoria with some notes on the diet of the host'. *Australian Veterinary Journal* 48: 456–61.
Coman, B.J. (1973). 'The diet of red foxes, *Vulpes vulpes* in Victoria'. *Australian Journal of Zoology* 21: 391–401.
Coman, B.J. & Brunner, H. (1972). 'Food habits of the feral house cat in Victoria'. *Journal of Wildlife Management* 36: 848–53.
Corbett, L.K. (1974), op. cit., Ch. 2.
Corbett, L.K. (1988). The identity and biology of Thai dingoes, op. cit., Ch. 3.
Corbett, L.K. (1989). 'Assessing the diet of dingoes from feces: a comparison of 3 methods', op. cit., Ch. 2.
Corbett, L.K. & Newsome, A.E. (1987). 'The feeding ecology of the dingo. III. Dietary relationships with widely fluctuating prey populations in arid Australia: an hypothesis of alternation of predation'. *Oecologia* 74:215–27.
Curio, E. (1976). *The Ethology of predation.* (Springer-Verlag: Berlin.)
Jarman, P.J. & Wright, S.M. (1993). 'Macropod studies at Wallaby Creek. IX. Exposure and responses of eastern grey kangaroos to dingoes'. *Wildlife Research* 20: 833–43.
Lunney, D., Triggs, B., Eby, P. & Ashby, B. (1990). 'Analysis of scats of dogs *Canis familiaris* and foxes *Vulpes vulpes* (Canidae: Carnivora) in coastal forests near Bega, New South Wales'. *Australian Wildlife Research* 17: 61–8.
Marsack, P. & Campbell, G. (1990). 'Feeding behaviour and diet of dingoes in the Nullarbor region, Western Australia'. *Australian Wildlife Research* 17: 349–57.
Newsome, A.E., Corbett, L.K., Catling, P.C. & Burt, R.J. (1983). 'The feeding ecology of the dingo. I. Stomach contents from trapping in south-eastern Australia, and the non-target wildlife also caught in dingo traps'. *Australian Wildlife Research* 10: 477–86.
Newsome, A.E. et al. (1983), op. cit., Ch. 4.
Robertshaw, J.D. & Harden, R.H. (1985). 'The ecology of the dingo in north-eastern New South Wales. II. Diet' and 'III. Analysis of macropod bone fragments found in dingo scats'. *Australian Wildlife Research* 12: 39–50, 163–71.
Robertshaw, J.D. & Harden, R.H. (1986). 'The ecology of the dingo in north-eastern New South Wales. IV. Prey selection by dingoes, and its effect on the major prey species, the swamp wallaby *Wallabia bicolor* (Desmarest)'. *Australian Wildlife Research* 13: 141–63.
Robertshaw, J.D. & Harden, R.H. (1989). 'Predation on Macropodoidea: a review'. *Kangaroos, Wallabies and Rat-kangaroos, Vol. 2.* (eds Grigg. G., Jarman, P. & Hume, I). (Surrey Beatty & Sons: New South Wales.) pp. 735–53.
Shepherd, N.C. (1981). 'Predation of red kangaroos, *Macropus rufus*, by the dingo, *Canis familiaris dingo* (Blumenbach), in north-western New South Wales'. *Australian Wildlife Research* 8: 255–62.
Stephens, D.W. & Krebs, J.R. (1986). *Foraging Theory.* (Princeton University Press: Princeton.)
Strahan, R. (ed.) (1983). *The Australian Museum Complete Book of Australian Mammals.* (Angus & Robertson: Sydney.)
Thomson, P.C. (1992). 'The behavioural ecology of dingoes in north-western Australia. III. Hunting and feeding behaviour, and diet'. *Wildlife Research* 19: 531–41.
Triggs, B., Brunner, H. & Cullen, J.M. (1984). 'The food of fox, dog and cat in Croajingalong National Park, south-eastern Victoria'. *Australian Wildlife Research* 11: 491–9.
Whitehouse, S.J.O. (1977). 'The diet of the dingo in Western Australia'. *Australian Wildlife Research* 4: 145–50.

Wright, S.M. (1993). 'Observations of the behaviour of male eastern grey kangaroos when attacked by dingoes'. *Australian Wildlife Research* 20: 845–9.

CHAPTER 8 Population dynamics

Bacon, P.J. (ed.) (1985). *Population Dynamics of Rabies in Wildlife*. (Academic Press: London.)

Baldock, F.C., Thompson, R.C.A., Kumaratilake, L.M. & Shield, J. (1985). '*Echinococcus granulosus* in farm dogs and dingoes in south-eastern Queensland'. *Australian Veterinary Journal* 62: 335–7.

Best, L.W., Corbett, L.K., Stephens, D.R. & Newsome, A.E. (1974). 'Baiting trials for dingoes in central Australia, with poison '1080', encapsulated strychnine, and strychnine suspended in methyl cellulose'. *CSIRO Division of Wildlife Research Technical Paper* 30: 1–7.

Breckwoldt, R. (1988). *A Very Elegant Animal The Dingo*. (Angus & Robertson Publishers: Sydney.)

Coman, B.J. (1972), op. cit., Ch. 7.

Corbett, L.K. (1974), op. cit., Ch. 2.

Dunsmore, J. & Burt, R. J. (1972). '*Filaroides osleri* in dingoes in south-eastern Australia'. *Australian Veterinary Journal* 48: 548–51.

Geering, W. A. & Forman, A. J. (1987). *Animal Health in Australia, Vol. 9, Exotic Diseases*. Bureau of Rural Science, Department of Primary Industries & Energy (Aust. Govt Pub. Service: Canberra.)

Jubb, K.V.F., Kennedy, P.C. & Palmer, N. (eds) (1993). *Pathology of Domestic Animals*, Vol. 1 (4th edn). (Academic Press, Inc: San Diego.)

McIlroy, J.C. et al., op. cit., Ch.4.

Newsome, A.E., Corbett, L.K. & Stephens, D.R. (1972). 'Assessment of an aerial baiting campaign against dingoes in central Australia'. *CSIRO Division of Wildlife Research Technical Paper* 24: 1–11.

Newsome, A. & Catling, P. (1992). 'Host range and its implications for wildlife rabies in Australia'. Bureau of Rural Resources Proceedings No. 11. pp. 97–107.

Newsome, A.E., Catling, P.C., Burt, R.J. & Corbett, L.K. (nd). '*Echinococcus granulosus* and other parasites of wild canids in south-eastern Australia'. Unpub. ms.

Rabies control in Asia. Draft Report of Symposium, Jakarta, April 1993. Foundation Marcel Merieux and World Health Organisation.

Seddon, H.R. (1965). *Diseases of Domestic Animals in Australia. Bacterial Diseases*. Vols I & II, Pt 5 (2nd edn). Dept of Health Service Publ. Nos 9, 10. (Commonwealth of Australia: Canberra.)

Seddon, H.R. (1966). *Diseases of Domestic Animals in Australia. Protozoan and Viral Diseases*. Pt 4 (2nd edn). Dept of Health Service Publ. No.8. (C'wealth of Aust: Canberra.)

Seddon, H.R. (1967). *Diseases of Domestic Animals in Australia. Helminth Infestations*, Pt 1 (2nd edn); *Arthropod Infestations (Flies, Lice and Fleas)*, Pt 2 (2nd edn). Dept of Health Service Publ. Nos 5, 6. (C'wealth of Aust: Canberra.)

Seddon, H.R. (1968). *Diseases of Domestic Animals in Australia. Arthropod Infestations (Ticks and Mites)*. Pt 3 (2nd edn). Dept of Health Service Publ. No. 7. (C'wealth of Aust: Canberra.)

Spencer, B. (ed.) (1896). *Report of the Work of the Horn Expedition to Central Australia. Part II. Zoology*, op. cit., Ch. 4.

Thompson, R.C.A. & Kumaratilake, L.M. (1985). 'Comparative development of Australian strains of *Echinococcus granulosus* in dingoes (*Canis familiaris dingo*) and domestic dogs (*C. f. familiaris*), with further evidence for the origin of the Australian sylvatic strain'. *International Journal of Parasitology* 15: 535–42.

Thomson, P.C. (1992). 'The behavioural ecology of dingoes in north-western Australia. I', op. cit., Ch. 3.

Thomson, P.C. et al. (1992). 'The behavioural ecology of dingoes in north-western Australia. V', op. cit., Ch. 3.

Thomson, P.C., Rose, K. & Kok, N.E. (1992). 'The behavioural ecology of dingoes in north-western Australia. VI. Temporary extraterritorial movements and dispersal'. *Wildlife Research* 19: 585–95.

CHAPTER 9 Predator–prey interactions

Bauer, F.H. (1983). 'The coming of European man'. *Man in the Centre* (ed. Crook, G.). Proceedings of a Symposium, Alice Springs, April 1979. (CSIRO: Melbourne.) pp. 26–45.

Caughley, G., Grigg, G.C., Caughley, J. & Hill, G.J.E. (1980). 'Does dingo predation control the densities of kangaroos and emus?' *Australian Wildlife Research* 7: 1–12.

Corbett, L.K. (1979).' Feeding ecology and social organisation of wildcats (*Felis silvestris*) and domestic cats (*Felis catus*) in Scotland'. Unpub. PhD Thesis, Aberdeen University.

Corbett, L.K. & Newsome, A.E. (1987). 'The feeding ecology of the dingo. III', op. cit., Ch. 7.

Corbett, L.K. (1989), op. cit., Ch. 2.

Corbett, L.K. (in press). 'Does dingo predation or buffalo competition regulate feral pig populations in the Australian wet-dry tropics?: an experimental study'. *Wildlife Research* 21.

Finlayson, H.H. (1935), op. cit., Ch. 1.

Flamholtz, C. (1986), op. cit., Ch. 1.

Foley, J.C. (1957). 'Droughts in Australia. Review of records from earliest years of settlement to 1955'. *Bureau of Meteorology Bulletin* 43. (Commonwealth of Australia: Melbourne.)

Friedel, M.H., Foran, B.D. & Stafford Smith, D.M. (1990). 'Where the creeks run dry or ten feet high: pastoral management in arid Australia'. *Proceedings of the Ecological Society of Australia* 16: 185–94.

Goodson, P. (1993). 'Maremmas – doing what comes naturally'. *Merigal*, July 1993, pp. 15–16.

Lundie-Jenkins, G., Corbett, L.K. & Phillips, C.M. (1993). 'Ecology of the rufous hare-wallaby, *Lagorchestes hirsutus* Gould (Marsupialia: Macropodidae) in the Tanami Desert, Northern Territory. III. Interactions with introduced mammal species'. *Wildlife Research* 20: 495–511.

Morton, S.R. (1990). 'The impact of European settlement on the vertebrate animals of arid Australia: a conceptual model'. *Proceedings of the Ecological Society of Australia* 16: 201–13.

Newsome, A.E. & Corbett, L.K. (1977). 'The effects of native, feral and domestic animals on the productivity of the Australian rangelands'. *The Impact of Herbivores on Arid and Semi-arid Rangelands* (ed. Australian Rangeland Society). (Australian Rangeland Society: Perth.) pp. 332–56.

Newsome, A.E. (1983). 'Native fauna as indicators of range condition'. *Man in the Centre* (ed. Crook, G.). Proceedings of a Symposium, Alice Springs, April 1979. (CSIRO: Melbourne.) pp. 124–42.

Newsome, A. (1990). 'The control of vertebrate pests by vertebrate predators'. *Trends in Ecology and Evolution* 5: 187–91.

Newsome, A.E. et al. (1983), op. cit., Ch. 4.

Newsome, A.E., Parer, I. & Catling, P.C. (1989). 'Prolonged prey suppression by carnivores — predator-removal experiments'. *Oecologia* 78: 458–67.

Pech, R.P., Sinclair, A.R.E., Newsome, A.E. & Catling, P.C. (1992). 'Limits to predator regulation of rabbits in Australia: evidence from predator-removal experiments'. *Oecologia* 89: 102–12.

Pemberton, D. & Renouf, D. (1993). 'A field study of communication and social behaviour of the Tasmanian devil at feeding sites'. *Australian Journal of Zoology* 41: 507–26.

Ridpath, M.G. (1991). 'Feral mammals and their environment'. *Monsoonal Australia: Landscape, Ecology and Man in the Northern Lowlands* (eds C.D. Haynes, M.G. Ridpath & M.A.J. Williams). (A.A. Balkema: Rotterdam.) pp. 169–91.

Robertshaw, J.D. & Harden, R.H. (1989). 'Predation on Macropodoidea: a review', op. cit., Ch. 7.

Sheldon, J.W. (1992), op. cit., Ch .3.

Shepherd, N.C. (1981). 'Predation of red kangaroos, *Macropus rufus*, by the dingo, *Canis familiaris dingo* (Blumenbach), in north-western New South Wales', op. cit., Ch. 7.

Sinclair, A.R.E. (1989). 'Population regulation in animals'. *Ecological Concepts* (ed. J. M. Cherrett). (Symposium of the British Ecological Society: London.) pp. 197–241.

Spencer, B. (ed.) (1896), op. cit., Ch. 4.

Thomson, P.C. (1992). 'The behavioural ecology of dingoes in north-western Australia. III.', op. cit., Ch. 7.

Thomson, P.C. et al. (1992). 'The behavioural ecology of dingoes in north-western Australia. VI', op. cit., Ch. 8.

Walker, B.H. & Noy-Meir, I. (1982). 'Aspects of the stability and resilience of savanna ecosystems'. *Ecology of Tropical Savannas* (eds B.J. Huntly & B.H. Walker). (Springer-Verlag.) pp. 556–90.

Woodall, P.F. (1983). 'Distribution and population dynamics of dingoes (*Canis familiaris*) and feral pigs (*Sus scrofa*) in Queensland, 1945–1976'. *Journal of Applied Ecology* 20: 85–95.

CHAPTER 10
The future of expatriate dingoes

Corbett, L.K. (1988), op. cit., Ch .3.

Corbett, L.K. (1985), op. cit., Ch. 1.

Corbett, L.K. (1974), op. cit., Ch. 2.

Gloyd, J.S. (1992). 'Wolf hybrids: a biological time bomb?' *Journal of the American Veterinary Medical Association* 201: 381–2.

Jones, E. (1990). 'Physical characteristics and taxonomic status of wild canids, *Canis familiaris*, from the eastern highlands of Victoria'. *Australian Wildlife Research* 17: 69–81.

Newsome, A.E., Corbett, L.K. & Carpenter, S.M. (1980). 'The identity of the dingo. I', op. cit., Ch. 3.

Newsome, A.E. & Corbett, L. (1982). 'The identity of the dingo. II', op. cit., Ch.3.

Newsome, A.E. & Corbett, L. (1985). 'The identity of the dingo. III', op. cit., Ch. 3.

Nerabamtung, C. (1983). *Siamese Ridgeback*. (Siamese Cat Books: Bangkok.)

APPENDIX A
Glossary of zoological and botanical species

CSIRO Division of Entomology (1970). *The Insects of Australia*. (Melbourne University Press: Melbourne.)

Ehmann, H. (1992). *Encyclopedia of Australian Animals. Reptiles*. (Angus & Robertson: Sydney.)

Groves, R.H. (ed.) (1981). *Australian Vegetation*. (Cambridge University Press: Cambridge.)

Lindsey, T.R. (1992). *Encyclopedia of Australian Animals. Birds*. (Angus and Robertson: Sydney.)

Strahan, R. (1992). *Encyclopedia of Australian Animals. Mammals.* (Angus & Robertson: Sydney.)
Walker, E.P. et al. (1968). *Mammals of the World.* vols I, II (2nd ed.). (The Johns Hopkins University Press: Baltimore.)
Wilson, S.K. & Knowles, D.G. (1988). *Australia's Reptiles.* (Collins: Sydney.)
Readers Digest Services (1986). *Readers Digest Complete Book of Australian Birds* (2nd edn). (Readers Digest: Sydney.)

Plate numbers refer to colour photographs